D0044953

OCEAN'S END

*Travels Through
Endangered Seas*

COLIN WOODARD

BASIC BOOKS

A Member of the Perseus Books Group

Published by Basic Books,
A Member of the Perseus Books Group

Designed by Rachel Hegarty
All maps were designed by Jojo Gragasin.

Library of Congress Cataloging-in-Publication Data
Woodard, Colin, 1968–
   Ocean's end : travels through endangered seas / Colin Woodard.
      p.   cm.
   Includes bibliographical references and index.
   ISBN 0-465-01571-9
   1. Marine pollution.   2. Marine resources conservation.   I. Title.
GC1085 .W66 2000
363.739'4'09162—dc21   99-051771
                                                                              CIP

01 02 03   10 9 8 7 6 5 4 3 2 1

For Mom

Contents

Foreword

C OLIN WOODARD AND I MET ON A CRUISE on the Black Sea in 1997. It was a wonderful place to learn about the problems of our oceans, and as this book demonstrates, Colin did. It is not a pretty story, but it is one everyone should know. The plight of fishers, from Atlantic Canada to the coral-ringed islands of the central Pacific, is our plight as well. When ecologists talk about the connections that bind the life-support systems of Earth together, they're not kidding. Humanity is dependent on the oceans for obvious things, from high-quality protein to recreation. But it is also dependent on marine systems for the relative stability of the world's climate—climate that we count on to allow us to grow enough food for an exploding population of more than six billion people.

The basic problem, of course, is the increased scale of the human enterprise. That does not just mean brute population growth (although there certainly has been enough of that). It's also an ever-increasing impact per person as consumption, especially among the rich, escalates. In the past 150 years, human impacts on the oceans have multiplied more than twenty-fold—about 5-fold because of population growth and about 4-fold due to increased consumption and the use of environmentally malignant technologies and sociopolitical arrangements.

Those impacts are as varied as they are dangerous. Many fisheries stocks are being harvested too heavily, which in itself can inflict more or less permanent damage on populations of economically valuable fishes, as well as on fishing communities. But often the harvesting process harms the marine environment. Trawlers, for instance, drag heavy nets over the ocean floor, in many areas more than once a year, which can destroy sponges, corals, bryozoans, and other bottom-dwelling organisms that feed on materials suspended in the water. The destruction of the ocean floor can alter oceanic ecosystems in ways that reduce fisheries productivity. Furthermore, as people increasingly crowd into seashore areas, they cause destruction of coastal wetlands that serve as nurseries for many oceanic fishes, including those sought for dinner tables. Nutrients from human sewage and agricultural run-off have already created oceanic "dead zones," areas near the bottom too lacking in oxygen to support fishes or shrimps. Among the best known are one in the Atlantic near New York City and one in the Gulf of Mexico, which reaches the size of the state of New Jersey or more in mid-summer.

Many of these problems can be ameliorated by controls on fisheries harvesting, coastal zoning, strong restrictions on environmentally destructive aquaculture operations, building sewage treatment plants, constraining releases from animal feeding operations, and the like. But there clearly is nothing resembling a permanent solution to the problems of the oceans without a gradual and humane reduction of human population size and a serious attempt to constrain runaway consumption. In principle we know how to do the former (and recent trends in some regions have been in the right direction); the latter is probably the most daunting challenge facing humanity. But only controlling these fundamental drivers will save the oceans and give *Homo sapiens* a reasonable chance of achieving a sustainable society.

Before we can take any steps toward this goal, however, we require the motivation to act. People need to learn the seriousness and multidimensional character of the human predicament itself. That's where *Ocean's End* will make an important contribution.

Paul R. Ehrlich

Preface

THIS BOOK BEGINS IN EASTERN EUROPE, which was my home for nearly five years. I lived longest in landlocked Hungary and traveled throughout the region writing about the collapse of one system, efforts to build a more promising replacement, and the scheming few who for personal advantage harnessed—and in some cases unleashed—the darkest demons of Europe's past. In academic papers and, later, newspapers and magazines, I often found myself exploring these themes through environmental issues.

This may seem an unusual approach, but in 1989 it made perfect sense. In the late 1980s, many of Eastern Europe's Communist regimes thought environmental issues were a safe, apolitical forum through which the public could be allowed to blow off some steam without threatening the existing order. They unwittingly wedged an opening that would help bring down one-party rule. The first mass protests in Hungary were organized against a hydroelectric project being built by Hungary and Czechoslovakia on the Danube River. In 1989, Bulgaria's Ecoglasnost—a group formed to protest air pollution in Ruse from a Romanian chemical plant—led the protests that allowed reform Communists to overthrow Todor Zhivkov's hard-line regime. By several accounts, the 1987 Chernobyl disaster prompted Mikhail Gorbachev to move faster and farther with Glasnost than he originally intended. By disclosing the details of this and other accidents and allowing the public to respond, Moscow opened a torrent of uncensored speech and triggered large environmental protests in Latvia, Armenia, and Ukraine, some with nationalist undertones.

My first environmental reporting dealt with a river that led me, with time, to the sea. Hungarians were outraged when newly independent Slovakia completed its half of the controversial Gabčikovo-Nagymaros hydroelectric project and used it to divert the Danube—the inter-state boundary—into Slovakia for several miles. The Danube is central to Hungary's identity, the subject of waltzes, poems, and paintings. Slovakia's ex-Communist leaders chose the Gabčikovo dam as the central symbol of their nation's reawakening. Relations between the two nations turned icy but, in the end, cooler heads prevailed.

The Danube, however, did not. The sewers of Bratislava, the Slovak capital, emptied directly into the river which, once diverted, no longer passed through the wetlands region that once cleansed it. At first I focused on the implications for Budapest's water supply, which came from the river. Later I became interested in the overall health of the river system, which was very poor. Across a wide swath of Europe, cities, fields, livestock pens, and factories dumped their wastes into the Danube, which conveyed the pollution into the Black Sea. Soon news services in Bulgaria, Romania, and the USSR carried reports of a shocking environmental catastrophe. Through a combination of long- and short-term stresses, humans had triggered the sudden, rapid destruction of life in an entire sea.

On vacation in Maine I heard the news that the Grand Banks, perhaps the world's most productive fishery, had been closed for lack of cod. The Georges Bank off New England soon followed. When I got back to Europe, violent conflicts between British and Spanish fishermen made front-page news; the root problem was that there weren't enough fish to go around. A Norwegian environmental group revealed that the USSR had dumped sixteen nuclear reactors, most with their spent fuel assemblies, into the Kara Sea. Pacific salmon, Atlantic bluefin tuna, and North Atlantic swordfish were now scarce. A college classmate said something off-handedly about how he thought people were killing the oceans. At the time I dismissed the idea as far-fetched, maybe even hysterical, yet it stuck in my mind, gnawing quietly on the edges of my attention. But back in the Balkans I had other, more immediate things to focus on.

In early 1997, I finished a stint covering the international community's operations in postwar Bosnia-Herzegovina and decided I'd finally had my fill of Balkan fascism and the response of the West. I returned home to Maine and, between short assignments overseas, found myself researching the state of the marine environment.

At first it was only to satisfy my own curiosity. I grew up on and around the sea and had felt its pull while living in landlocked Budapest, Zagreb, and Sarajevo; if something systemic were happening to the northwest Atlantic or the wider ocean, I guess I wanted to know about it sooner rather than later. The more I read, the more scientists I talked to, the clearer it became that the oceans were in serious trouble.

So I did what I usually do. I packed my bags and headed off to see for myself. For a year and a half I crisscrossed the world ocean, from the foggy North Atlantic to fecund bayous, from balmy island shores in the Caribbean and equatorial Pacific to the frigid, glaciated coasts of Antarctica. I talked with fishermen and scientists, officials and activists, divers and sailors, religious missionaries and government ministers. In the end, I hid away in cabins on a research ship, the coast of Maine, and the central arteries of the nation's capital to write this book.

Ocean's End begins, as did my own journey, on the Black Sea, whose near-total destruction provides a cautionary tale of what can happen when marine environments are treated with reckless abandon. Chapter 2 provides an overview of the oceans, their central role in the story of life on Earth, and the crises they confront.

People tend to think of the crises in the oceans almost entirely in terms of fish and fisheries, and chapter 3 introduces these problems, describing the sad destruction of Grand Banks cod and, in the process, the basis of Newfoundland society. But fishing is only a small part of the picture. Chapter 4 deals with the interrelationship of land and sea and shows how the reengineering of the Mississippi River basin triggered enormous ecological disruptions not only in the Gulf of Mexico but throughout Louisiana's bayou country. For chapter 5 I traveled to Belize to investigate the decline of coral reefs and to dive with scientists on both damaged and pristine reefs.

Global change dominates chapters 6 and 7. The first covers a trip to the Republic of the Marshall Islands, one of several nations that face total destruction in the face of rising seas. The latter describes my journey to the Antarctic Peninsula, which lies on the front lines of global warming and ozone depletion.

The final chapter offers some practical, real-world strategies for remedying our dysfunctional relationship with the oceans, integrating the ideas of many people who have devoted their lives to marine resource management. The political and social challenges to adopting these measures are significant, but nowhere near as huge as the emerging crisis beneath the waves. We've proved at least two age-old adages to be incorrect: The oceans are not endless, and we can't safely assume that there will always be other fish within them.

Washington, DC
July 1999

OCEAN'S
END

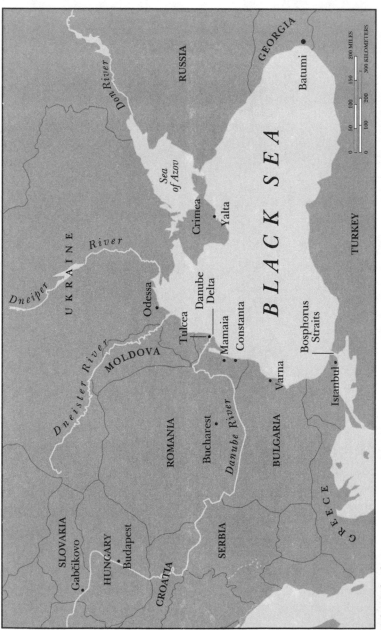

The Black Sea

one

Dead Seas

WALK OUT ONTO THE UGLY PIERS that line the prole-
tarian bathing strands of Yalta and look into the water.
You'll see them everywhere. Ubiquitous jellies of assorted shapes and sizes rising and falling in the rollers
under the mountainous Crimean shores: undulating round jellyfish,
ovoid comb jellies, finger-sized transparent creatures that feed on
the specks of life that live in each drop of sea water.

Under your feet, bag-like comb jellies move about like automated
vacuum cleaners eating anything in their path. There's one near the
piling you're standing beside. It looks extraterrestrial, and it is in
fact alien to the Black Sea. One end of the creature opens wide; its
whole body convulses, refracting sunlight in a shimmering rainbow
as it sucks everything in the vicinity into its formless insides. Larval
clams, fish, and crabs, hapless juvenile moon jellyfish, and microscopic copepods—whole generations are falling before this endless
alien herd.

You can see them from every pier and beach, each time you look
over your ship's railing or wade in the murky surf. From the rainy
shores of eastern Turkey to the dry headlands of Yalta you see them in
countless multitudes, dotting the Black Sea like stars in the night sky.

They've grazed the sea nearly clean, these voracious comb jellies.
Their numbers are unthinkably huge: a billion tons at last estimate,
more than ten times the weight of all the fish landed by all the fishermen in the world in a year. In a few short years they've all but con-

quered this ancient sea, starving out fish and dolphins, emptying fishermen's nets, and converting the web of life into brainless, wraith-like blobs of jelly.

∞

THE BLACK SEA IS NEARLY LANDLOCKED, the most isolated antechamber of the world ocean. Its only outlet is through the narrow Bosporus and Dardanelles Straits in Turkey, and these lead into the Mediterranean, which is itself connected to the Atlantic by a shallow, narrow strait. In some places, the Dardanelles and Bosporus are only 0.4 miles wide, and it takes 150 years for the Black Sea to replenish itself completely through these channels.

The Black Sea is a large, kidney-shaped body of water, 160,000 square miles, the same size as California. And it is deep—seven thousand feet in some places; it contains several times more water than the Great Lakes combined. Unique to the world ocean, the depths of the Black Sea do not support life and haven't for many thousands of years. Below a certain depth there is no oxygen in the salty water. All life lives in the sea's uppermost layers, a few hundred yards from the surface. Until very recently, life concentrated on a shallow shelf in the sea's northwest lobe between the low Danube Delta and the mountainous Crimean peninsula.

Humans have also concentrated near its shores, drawn to this and other sheltered seas by milder climates, bountiful food supplies, and an easy means of trade. Ancient Greece was built upon fish and grain supplies from its Black Sea colonies, and these supplies also sustained Byzantium and Imperial Russia, the Ottoman Empire, and the Soviet Union. Today six countries share its shores: Turkey to the south, Georgia and Russia to the east, Ukraine on the north, Romania and Bulgaria to the west. The waters of another dozen countries drain here via four massive river systems: the Danube, Dneister, Dneiper, and Don. These countries differ in culture, history, politics, and wealth, but they all share the Black Sea drainage basin. The activities of political and economic planners in Central Europe have serious implications for communities like Trabzon,

Turkey, or Batumi, Georgia, located a thousand miles away in northwest Asia. Water, the world's most precious resource, respects gravity rather than political frontiers.

Many of the world's seas suffer from some or all of the same symptoms as the Black, but they cannot be understood in isolation from one another. The experience of the Black Sea is a warning of what is to come for other sheltered seas, indeed for the oceans at large. It shows that the seas will not survive if people continue to regard the oceans as a bottomless sewer into which wastes of all sorts can be endlessly deposited without ill effect, while we continuously extract food and other resources in ever-greater quantities. Everything has limits, and the world ocean is signaling that its carrying capacity has been exceeded. Human-caused stresses are weakening bays and harbors, gulfs and seas, coastal shelves and offshore fishing banks all around us. Marine scientists are frankly frightened by the scale and complexity of the emerging marine crisis. It is one that can only be averted if there is a dramatic change in our understanding of the ocean and its central role in maintaining life on Earth.

The Black Sea is a microcosm of what is happening to the ocean systems at large. The stresses piled up: overfishing, oil spills, industrial discharges, nutrient pollution, wetlands destruction, the introduction of an alien species. The sea weakened, slowly at first, then collapsed with shocking suddenness. The lessons of this tragedy should not be lost to the rest of us, because much of what happened here is being repeated all over the world. The ecological stresses imposed on the Black Sea were not unique to communism. Nor, sadly, was the failure of governments to respond to the emerging crisis.

❧

BATUMI, GEORGIA, FALL 1997

It was a torrential downpour, a vertical blast of rain driven forth from the crevices of the Caucasus high above. My umbrella turned inside out as soon as I left the mud-splattered bus. I was soaked through by the time I reached the derelict concrete bridge spanning the nearby river.

I had traveled for four days by car, aircraft, train, ship, and bus to find myself walking through a muddy garbage dump outside Batumi, the largest port in the former Soviet republic of Georgia. The city stands on the far eastern end of the Black Sea, where Europe and Asia meet on a bit of Near Eastern coastline at the corner of Turkey and Georgia. It's part of a country that's home to one of the oldest civilizations on Earth, tenaciously surviving alongside more powerful neighbors and pushing aside would-be invaders. Georgians say that when God divided the Earth among its peoples, the Georgians were sleeping after a long night of partying. By the time they finally showed up, God was out of land. But He had enjoyed their party so much that He gave them the bit of land He had been saving for himself: a wedge of warm beaches, green river gorges, and rugged snow-capped mountains. Unfortunately, God hasn't been so generous to Georgians since.

The dumping ground was littered with old Christmas tinsel, rusty cans, a cheap aluminum fork, and the plastic torso of a discarded doll. Chunks of machinery and vehicles were strewn among corroding storage drums. A pair of grazing cows were feeding on weeds growing on top of a decaying Russian bulldozer. The dump's orange-vested attendants, seemingly embarrassed that foreigners had seen the livestock here, encouraged the beasts to leave by throwing rocks at them. The cows retreated into a garbage-strewn depression nearby to continue their feeding.

From the bridge I could see racing gray water through tractor-sized holes in the roadway. This Soviet-era solid waste dump was eroding into the river, the rain-soaked soil sloughing off the embankment as we watched. Oil, battery acid, and whatever was in the storage drums was seeping into the watershed, which emptied into the Black Sea. This had been going on for years.

Fortunately the problem had the attention of the government of newly independent Georgia. In fact, Deputy Environment Minister Tengiz Gordeladze had brought us here and was now standing in our midst and yelling at the top of his lungs so as to be heard over the gale.

"We need help from the outside world," Gordeladze was shouting over the howling wind. His suit was completely soaked and rain streamed off the end of his nose. "Something has to be done."

It rained all that day in Batumi, which had been the rainiest place in the USSR when there was such a country. In town, green algae grew along the damp walls of buildings. Beggars took shelter under concrete awnings. Water puddled in the broken pavements until a passing vehicle splashed it out onto unlucky pedestrians. The mountains remained shrouded in mist. French sailors in the last century called Batumi "Le pissoir de la Mer Noire" because of the rain. These days this could apply equally well to the untreated human sewage pouring into the sea from rivers and sewer drains.

The main street was a mix of old and new. Drab state-owned stores stood their ground amongst little new boutiques selling cheap goods brought across the nearby border with Turkey. It was difficult to tell which stores were open because nobody turned the lights on until well after dark. A war refugee was squatting in the doorway of the bookstore—an old man wearing a dirty, worn suit who'd repaired the frames of his thick glasses with masking tape. The cardboard box in front of him was covered in plastic to protect it from the rain. I noticed it was full of bank notes. "Change money?" he asked with little enthusiasm.

The old Soviet guidebook I purchased inside showed pictures of happy tourists packed onto palm-tree-lined beaches. A smiling young couple shared a bottle of wine in a hotel restaurant, he in a gray suit and sideburns, she wearing a burnt orange dress and lots of makeup. Concrete hotel towers stood behind a sunlit beach in a picture captioned "Contemporary architectural achievements ensure comfort!"

Outside we found some teenagers standing under umbrellas on the pebbly beach on the way out of town. Josef and Costa, the photographers I was with, took their pictures and invited them to have a coffee at a stand-up cafe. They turned out to be refugees from Abkhazia, a region the size of Long Island that had seceded from the rest of Georgia in a bloody war. They lived ten-to-a-room in one of the nearby hotels, now a refugee hostel. Only one had a parent who was employed. Another had a brother who worked on commission for a Turkish "businessman" changing money on the street—a popular occupation, judging from the number of moneychangers hanging around the main market.

We ate tasty potato dumplings under florescent lights in a worker's cafeteria. Several tables had been drawn together to allow a second grade class to celebrate their classmate's birthday. The children wore homemade sweaters, clean button-down shirts, and wool jackets. Two men at the next table spooned soup silently, occasionally staring at us with blank faces and sad, brown Georgian eyes. Josef included their table in the round of beers he ordered in his native Czech. They solemnly toasted him when they were served. "To your health," Josef suggested.

"Better to drink to the children," the older one said in Russian. "For us it is too late."

∞

CONSTANŢA, ROMANIA, SUMMER 1990

Batumi had been popular because it was the southernmost resort in the old Soviet Union, but for decades the resorts of the Romanian coast were the most coveted destination in the entire Eastern Bloc. The industrial working class packed Romania's beach resorts on their annual break from the gray factories and coal mines of Britain, Bulgaria, and Belorussia. They came by the millions each summer to bask on fine sand beaches and while away the nights over bottles of Dobrudgean wine and high-octane Tsuica in the open-air restaurants lining the shore. Two weeks at a Black Sea resort was one of the great perks afforded to the proletariat, and in the late 1970s Romania was the most prized destination. Those were heady days for the Romanian socialist utopia, when the hardest-working laborers of the socialist world came together in the hive-like concrete vacation complexes of Constanţa County for a party that lasted all summer long.

In the 1970s, when Sorin Strutinsky was a boy, the waters of the Black Sea were still clean and clear. His family lived in Constanţa, the ancient Greek port of Tomis, which for thirty years served as the main staging ground for the summer tourist invasion. For Constanţa residents it was easy to take in the mud cures at Eforie Nord or whisk the children off for a day on the beaches of Saturn, Neptune, Jupiter, or Venus, a chain of high-rise beachfront resorts stretching south to the Bulgarian border. It was a childhood spent in the midst of Lenin's "Daytona Beach" among topless women, naked, screaming children, and tipsy Ukrainian machinists.

One summer, while Sorin was still in high school, the sea began to change. The water had a sticky, slimy feel. It smelled bad—like rotting eggs—and there were lots of dead fish in it. After a swim in the Black Sea, bathers found their bodies covered in a strange coating. Afterwards, some became sick, and others stopped swimming in the sea altogether.

By the time of my first visit to the Romanian coast, the party was definitely over. It was late June 1990, and weeds surrounded the high-rise concrete hotels that stand shoulder-to-shoulder along the thin sandy isthmus of Mamaia, Romania's answer to Miami Beach. The hotels were all owned by the Ministry of Tourism and their dreary employees sat in the lobbies watching reruns of *Dallas*, which was the rage in Romania that summer. There was very little traffic along the strip, no cars or buses in the parking lots. Restaurants and bars had little more to offer than domestic brandy and greasy pork cutlets—there was no beer, wine, juice, or even mineral water. Nor were there any tourists to serve them to.

Sorin was then twenty-two and the editor-in-chief of an independent newspaper started by his university friends in the aftermath of that winter's bloody revolution. They'd watched television images of the fighting in Bucharest, where huge crowds braved freezing cold, tanks, and sniper fire to ensure the end of a dictatorship that had destroyed their lives. "We celebrated for a week," he recalled. "Until we realized that the new government was being run by the same people who ran the last one. Nicolae and Elena Ceauşescu are dead, but democracy's yet to be born."

So they'd started a newspaper in a rented basement and that summer had taken to denouncing the new government and concocting conspiracy theories as to what had really occurred that winter. When I came by their musty office to talk about it they piled everyone into an aging taxi and sped up to Mamaia for "lunch." This turned out to be an all-day affair, a procession of pork cutlets and watery brandy that I was usually appointed to pay for. We wandered from one restaurant to another, although they all served the same meal prepared exactly the same way, served on ubiquitous Romanian Tourist Office dishware beside aluminum forks on soiled, mustard yellow tablecloths. The entire town—all ninety high-rises with their accompanying restaurants, bars, and shops—had been built en masse in 1960–1963 following a Communist Party decree. It was really one gigantic hotel.

I didn't learn much about the revolution—Sorin's conspiracy theories were crazed and contradictory when he was sober and incomprehensible after several glasses of brandy-and-coke. Instead I watched one of his friends dive fully clothed into a brackish green swimming pool. When he surfaced he began singing along with the synthesizer-crazed band out on the patio. The tune was Abba's "Dancing Queen" but the lyrics were unintelligible.

"It's sad to see my town like this," Sorin said, leaning back in his chair to gaze up at fifteen stories of darkened hotel windows. "It's never going to be the same—and I don't mean just because of the end of communism. The sea has also changed, it's dirtier and uglier than before."

He paused and smiled wryly. "And so are we."

The next day I ditched Sorin and company and explored Mamaia on my own.

The beach was deserted except for some local teens rough-housing in the surf and a young East German family lying burnt-red under the afternoon sun. There were a few dead fish on the beach. Up the coast the Mediora Voda Chemical Plant could be seen spewing yellow smoke into the air and who-knows-what into the water. But when in Romania . . .

The water was cool and refreshing and I swam submerged out from the beach. Then I floated lazily on my back, semi-weightless like an unborn child. After a few minutes I noticed that I had drifted into an enormous slick of brown-tinged scum. The water was viscous and covered with bits of unhealthy brown foam. Algae mixed with human sewage, a slick the size of a football field stretching along the beachfront twenty yards from the water's edge. It was my first and last swim in the Black Sea.

Other tourists weren't coming to the Black Sea coast at all. Traian Arhip, the managing director of the state firm that operated Mamaia's strip of state-owned hotels, lamented the collapse of one of his country's most important industries. "Arrivals are off by ninety percent," he told me in his hectic Bucharest office. Hot, exhaust-laden air drifted into the third-story window from the busy boulevard below. A faded poster of a rather intimidating concrete resort complex hung on the drab wall. "The British made up more than half of our business before, but now they won't come on account of the shortages and political instability. The Soviets don't have any money, and the East Germans and Czechs are off seeing the West for the first time. It's a real disaster."

Romania's economic and political crisis had scared its tourists away. But on my future visits to the Romanian and Bulgarian coast I learned that Black Sea resort operators were discovering a new problem in winning tourists back. The sea that had drawn these herds of cash cows to their shores was now scaring them away. Fecal contamination and toxic algae blooms forced regular beach closures at the height of the short summer tourist season. There were outbreaks of cholera and hepatitis. The wind carried blankets of decomposing fish onto shore, to the consternation of vacationers and authorities alike.

"Repeat business," Arhip said dryly, "is way off."

❧

IT'S IRONIC THAT THE SEA IS BRINGING DEATH and malaise to Constanţa, Batumi, Varna, and other ancient cities. For it was the sea that nourished the civilizations that grew upon its shores—the Greeks and Scythians, Ottomans and Russians, modern Turkey and

Ukraine. It gave them food, moderated the climate, transported their cargo, and absorbed their wastes.

The Ancient Greeks named this the Euxine, or hospitable, Sea. Jason and his Argonauts came here in pursuit of the Golden Fleece. They discovered this enormous sea at the other side of a treacherous, narrow, and previously unknown strait, the Bosporus, where a strong current runs towards the Mediterranean. Running against the current in tiny Greek coastal craft was no easy feat, and the Argonauts' Bronze Age journey would not be repeated until the eighth century B.C.

Emerging from the long, difficult passage through the straits, the Greeks found fish and timber in great abundance. On the shores of the Crimea and what is now Russia and Ukraine the colonists planted wheat. Fish, grain, and lumber—three resources the city-states of Hellenic Greece had long since exhausted—were plentiful around these new colonies, including Tomis (Constanţa). Independent traders kept Greek civilization going, delivering the bounty of this most Euxine region.

Like the civilizations that followed them, the Greeks' survival depended in large part on the healthy functioning of vast communities of life on the shore and within the salty depths. The Euxine Greeks caught and ate fish, cured them, and traded them with Hellenic merchants in exchange for olive oil and manufactured goods from Athens, Egypt, and beyond. The hungry city-states of the Aegean were exhausting the arable lands outside their walls and increasingly relied on the Black Sea trade for their sustenance and for lumber to build the ships that maintained these lifelines. Later they planted wheat on the great Russian steppes, the first seeds of what would remain the breadbasket of the Western world for almost three thousand years.

But it was the creatures of the Black Sea that brought human civilization here and allowed it to prosper on its shores. In the lagoons at the mouths of the great rivers one could easily net pike, salmon, and shad. Schools of enormous sturgeon rushed into the estuaries to breed, and the females were harvested for the huge quantities of black caviar found within; these fish eggs were so plentiful that in fourteenth-century Byzantium caviar was the food of the poor. So productive were the sturgeon of the Danube, Dneiper, and Dneister estuaries that today the words "Russian" and "caviar" are linked in the popular mind.

Schools of anchovies circumnavigated the sea each year. So timely were their migrations that fishermen could predict their annual arrival near their home ports within a day or two. These enormous schools fattened themselves on the northwest shelf, then circled the south coast, consuming tiny floating animals, and wintered in the warmer waters off Batumi and Sukhumi. They were pursued all the while by bonito, mackerel, tuna, and dolphins, part of a food chain topped by fishermen seeking to net them all. The fishermen's job was made easier by the massive springtime migration through the narrow Kerch Straits into the spawning grounds of the Sea of Azov. One had only to stand on the shore here or at the entrance of other bays on the northern coast to catch as many nets full of fish as one could carry. Farmers spread anchovies on their fields as cheap fertilizer, and fishing ports around the sea prospered with each harvest of the great migrating herd.

The anchovies, their predators, and the fishermen all depended on the remarkable sea-grass meadows and kelp forests of the northwest shelf. Inside any bay or inlet that afforded protection from the waves grew great submarine pastures of *Zostera* grass, which sheltered a cornucopia of life: amphipods, isopods, and shrimp; mollusks, crabs, and pipefish; feasting grounds for sprat, turbot, mullet, whiting, and rays. These or their larvae provided food for open-water fish like the anchovy; many species were delicacies for humans as well.

As the *Zostera* meadows petered out in deeper water, gravelly banks of broken mussel shells sheltered huge quantities of pearl-producing oysters. Still farther out, the sea was ringed with a thick carpet of sea mussels. These flourished in the Black Sea in unusually large numbers, growing fat by filtering the rich nutrients emptied into the sea by the great rivers of Europe.

But the center of this vast ecological web lay still farther offshore. In the middle of the shelf grew an enormous forest of brownish-red *Phyllophora* kelp, a rooted cousin of the famous sargassum that can grow to a height of twenty feet. Here it covered fifty-eight hundred square miles and had an aggregate mass of as much as six million metric tons—the largest concentration of rooted red algae in the entire world ocean. This forest was one of the keystones of the Black Sea's living community; it fed and sheltered 170 animal species—sponges, anemones, shrimp, and crabs—many of them found only in this body of water. It also was a principal source of oxygen on the northwest shelf, where most marine life is found. Yuvenaly Zaitsev of the Ukrainian Academy of Sciences estimates that it produced two million cubic meters of oxygen a day, which is enough to fill three Superdomes with pure oxygen. Humans harvested the tops of the kelp forests for agar, the stuff that encapsulates your vitamin E pills; keeps pre-packaged coffee cakes soft for weeks at a time; and assists in the healing of burns, the preservation of canned meats, and the cultivation of laboratory bacteria.

∞

ALL THIS IS GONE. The *Phyllophora* forests have died, taking their residents with them. The mussel fields and oyster banks have disappeared, as have shrimp and crabs. The anchovy herds are all but destroyed. Wild sturgeon are endangered, the monk seal is probably extinct. It's become a most inhospitable sea for any creature more complex than a floating jelly.

This disaster was caused by the ignorance and indifference of countless people. Austrian farmers and Turkish fishermen, Serbian, Slovak, and Romanian hydro-engineers, Bulgarian hotel developers and Soviet party planners, oil company executives and shipping magnates—all share blame.

The sea endured many stresses, but the worst were the increasingly poisonous injections it received from its own tributaries. Four major rivers empty into the Black Sea and its companion, the Sea of Azov. One, the Danube, accounts for more than half of the total flow of fresh water and pollution into the sea. It is there, hundreds of miles from the Romanian coast, that one must begin looking for the causes of the Black Sea's ecological collapse.

∞

NORTHWEST HUNGARY, SPRING 1993

One day in 1993, the Hungarian Foreign Ministry rounded up a busload of foreign journalists and drove us 100 miles out of Budapest, up the Danube toward Slovakia. From Mosonmagyaróvár, the big Mercedes coach crept down tree-lined country lanes with an entirely unnecessary police escort. When we took a left onto a potholed dirt track, the Indian journalist sitting next to me wondered aloud if they were "finally going to get rid of the whole troublesome lot of us."

Tempting though it may have been, the Foreign Ministry was true to its word. After much bouncing and tousling we pulled into a scrubby clearing and up to the banks of the Danube. Or what had been the bank of the river's main arm until a few weeks earlier. For the Danube was no longer there. Some fifteen feet down the muddy bank a little brackish water covered the very bottom of this great river. Elsewhere many lesser arms had dried up altogether, trapping fish and other creatures in isolated scattered pools. Those that survived in the larger holes would die when their confined habitats froze solid that winter.

"The river," a ministry spokesman announced, "has been stolen."

We were in the midst of the Szigetköz, a vast wetland forest just inside the Hungarian border where the great river breaks into a peculiar inland delta of sidearms and rivulets a dozen miles long. It had been the richest fish habitat in all of Hungary, with more than sixty-five species. But its role in the protection of the river itself was even more important. Dense vegetative growth around us acted as a giant filter, cleansing the Danube more effectively than a hundred-million-dollar water treatment plant could. Entering the Szigetköz, the river was fouled with the untreated human sewage of Bratislava's half million residents. After passing through the Szigetköz the water was clean enough for agricultural, industrial, and (with a minimum of treatment) even household use.

Less than a mile upstream from the Hungarian frontier, the Slovaks recently had completed a massive diversion weir, creating an enormous new reservoir in the process. Instead of flowing into

Hungary and the Szigetköz, the Danube was channeled into a titanic artificial canal constructed to convey it across the flat agricultural fields of south Slovakia. There it was forced through turbines housed in the gray concrete Gabčikovo hydroelectric barrage before being released back into the original riverbed fifteen miles from where it was "borrowed." To accomplish this feat, Communist Czechoslovakia and neo-Communist Slovakia built a concrete-lined canal standing *above* the surface of the plain, hemming in Europe's second-largest river between huge earthen embankments. Rumor had it that the builders had rushed the canal to completion using substandard materials, an allegation supported by frustrated construction workers at the site. The director of Hydrostav, the firm that built the dam, insisted there were no serious engineering problems. Not surprisingly, the government backed him up; the director was the brother of Slovak Prime Minister Jan Čarnogursky.

I stood on what had been a country road linking two Slovak farming villages. Across the fields behind me I could see one of the villages. Twenty yards ahead the road was buried under a three-story-high wall of earth. Of the village beyond I could glimpse only the very top of a church spire poking up over the Danube's new conduit. The residents of nearby Dunajská Streda joke that they've started building an ark to deliver them from the deluge that would follow a rupture of the canal.

From the top of the dam itself the Danube looked as gray and lifeless as its bare concrete conduit. Unfiltered by the marshes, the river rushed towards the Black Sea, receiving runoff from assembly-line style meat-production pens and fields carelessly treated with the kind of potent fertilizers and pesticides that prompted Rachel Carson to write *Silent Spring*. Huge factories and power plants released heated water, petroleum, PCBs, dioxin, heavy metals, and acidic fallout into the river. The toilets and dishwater of eighty million people flush into the Danube with little or no treatment.

The Danube drains half a continent, from the alpine streams of the Swiss Alps and Black Forest to the entire river systems of Hungary, Slovakia, Romania, Bulgaria, and Serbia. In this densely populated part of the world, many of the unwanted byproducts of human endeavor wind up in the river. Until a century ago the river could cope with this burden reasonably well. Farmers spread dung on their fields and kept livestock in pastures, industrial enterprises released nothing more toxic than chlorine and sulfuric acid, and electrical power plants didn't exist.

Today the Danube, Dneister, Dneiper, and the Black Sea's other tributaries are the dumping grounds for large quantities of oil and pesticides, toxic industrial effluents, and radioactive material. The Industrial Age brought a dramatic and unprecedented growth in this region's production of wastes, with an ever-greater proportion of complex inorganic substances that nature has a difficult time digesting. Perhaps most damaging for aquatic life has been our own species' regular, prodigious output of organic wastes.

Infants who drink the water in Mezötur sometimes turn blue and die. This Hungarian farming community, like thousands of other villages on the Danube floodplains, is surrounded by huge single-crop fields of corn and soy, which are fed chemical fertilizers. Large quantities of nitrate residue have soaked into the water table underneath. The water tastes awful, to be sure, but it can also kill the weak by reducing the oxygen-carrying capacity of their blood. Drinking water has to be trucked in from towns with water treatment facilities. Abundant local water is used primarily to irrigate the surrounding fields and, so tainted, it makes its way into the river to wreak havoc of a different sort.

Microscopic specks of plant life live in every drop of surface water. Fed nutrients, they multiply rapidly. Nitrogen and phosphorus, their favorite foods, are present in great quantities in the manure, sewage, detergents, and fertilizers we drain into watersheds. Overfed, these tiny algae explode in enormous blooms, far outstripping the ability of marine animals to graze them down. Successive generations of short-lived algae die and sink to the bottom, where bacteria set to work decomposing them. In the process, the bacteria gobble up most or all of the oxygen in the surrounding bottom water.

When the amount of dissolved oxygen in the water falls below a certain level, fish, mollusks, crabs, and other creatures literally suffocate, their gills unable to capture enough oxygen to keep them alive. Sometimes a dead zone of oxygen-less water rises nearly to the surface of a pond, lake, or slow-moving stretch of river, killing almost everything in its path. In extreme cases, algae blooms form mats near the water's surface so thick that they block sunlight from reaching plants growing on the bottom, killing them along with the oxygen-breathing fish and animals. This process is called "eutrophication," and it affects lakes and rivers around the world.

Only very recently have we managed to kill an entire sea in such fashion.

Nutrient pollution in the Danube basin exploded as Eastern Europe became increasingly urbanized and its massive state farms and agricultural cooperatives began applying ever-greater quantities of chemical fertilizers to the land. The Lower Danube's phosphorus load increased by two and a half times between 1960 and 1990, while its organic nitrogen burden quadrupled. The river carried these nutrients downstream and released them into the Black Sea, where they produced some of the most serious and destructive algae explosions ever recorded—worse, even, than the dramatic growth in nutrients flows could account for.

Not until 1996 did scientists realize that the dams built along the Danube had made the algae blooms far more potent and destructive.

After the Danube passes through Budapest it runs south across the flat Hungarian plain. It passes through the shattered Croatian city of Vukovar and the untreated sewer drains of Belgrade and eventually takes a precipitous fall down the continental shelf to the low-lying plains of the central Balkans. This is a dangerous, turbulent stretch that had long been a hazard to navigation. In the twentieth century engineers recognized that if harnessed here the racing river could generate a great deal of electricity. Their dream was realized in 1972 with the completion of the Iron Gates Dam project. Yugoslavia and Romania—which built the project together on their common frontier—celebrated their technological and political achievement, which reduced their dependence on the Soviet Union for energy. They did not realize the extent to which they had altered the chemistry of a major natural system.

Before the dam's completion, alpine streams washed silica from the mountains into the Danube, which carried it to the sea. The presence of silica ensured that when explosive algae blooms occurred they would contain a large proportion of diatoms, tiny one-celled plants that live inside beautiful glass shells. Diatoms, like dinoflagellates and other algae bloom species, feed and reproduce in waters polluted by nitrogen and phosphorus found in human and animal wastes and artificial fertilizers; the difference is that they're not nearly as toxic and don't consume as much oxygen as their algal cousins. But the dams blocked most of the silica from reaching the sea, collecting it uselessly in artificial reservoirs; diatoms had more difficulty obtaining silica with which to grow their characteristic glass shells. Nitrogen and phosphorus arrived in great quantities from the cultivated floodplains of Romania and Bulgaria. By 1997, diatoms had bloomed to two and a half times their mid-1970s level, but dinoflagellates had increased six-fold. The result was an algae bloom that rapidly depleted oxygen from the water and was more likely to kill or poison fish and shellfish.

❧

DANUBE DELTA, ROMANIA, FALL 1997

As it approaches the sea, the Danube breaks into a maze of tendrils that thread through a two-million-acre wetland created by centuries of sedimentation. Similar to Hungary's Szigetköz wetlands but far larger, the Danube Delta was nature's ultimate insurance policy against pollution. When healthy, it acted as an enormous living filter, absorbing nutrients and toxins from the river before they reached the sea.

The delta officially begins in Tulcea, Romania, where the Danube breaks into three large and five smaller arms that meander toward the Black Sea. By the time it gets here, the river has run a course of seventeen hundred miles from its source in the Black Forest and has received the flow of 120 smaller rivers. From the Tulcea waterfront you see not the Blue Danube that inspired Strauss but rather a swampy shipping channel overshadowed by an aluminum smelting plant that spews foul smoke from its stacks.

The locally produced tourist brochure promises "a town that lives [in] the effervescence of development" due to the ongoing accomplishment of "new industrial objectives." Listed for the benefit of visitors, these include the aluminum smelter, an iron and steel mill, a shipyard, cannery, fruit factory, slaughter house, plastics plant, and brewery. "The town is being radically changed not only in the economic aspect, but also from the social viewpoint," the guide concludes cheerfully. I had to agree.

The town's residents lived in crumbling socialist apartment blocks along the river, built en masse less than a decade ago, when this was to be the staging ground for the draining and colonization of the wetlands. Torrents of human filth poured from open sewers right under the docks, where aging excursion boats stood by to pick up "ecotourists." The tourists stood awkwardly on the landing amid decaying cement buildings and rusting cargo barges that lined the trash-strewn embankment.

Out on the water it wasn't much better. The river reeked of oil and petroleum products, hardly surprising considering that 53,000 tons are spilled into it each year. Our boat was confined to an ugly shipping channel that sliced through sand dunes and wetland forests to the Black Sea, releasing the river's filth over the northwest shelf just north of the Romanian beaches. Wildlife avoided this and the other main channels, which have all been artificially straightened. Instead of pelicans and cranes, we saw great lines of five-story-tall gravel dredges rusting on their barges like a herd of mechanical dinosaurs. Around the next bend somebody had set up a small floating hotel for the ecotourists, powerboats and jet-skis tied up to its landing.

To Ceauşescu's thinking the rivers and lakes, reed swamps, meadows, and wetland forests of the Danube Delta represented a waste of Romania's productive potential. He ordered that the delta be drained, colonized, and planted with wheat, corn, and rice.

Hundreds of ring-shaped dikes were constructed, and the area within each was drained, cleared, and planted with crops. This disrupted the hydrological balance, allowing sea water to intrude into the soil under the new fields, killing the crops. Although failure was inevitable, Ceauşescu pushed ahead with his Complex Program for the Intensive Exploitation of the Resources of the Delta. Fortunately for the delta, Ceauşescu's other, equally insane development policies so undermined Romania's economy and society that the Complex Program was never completed. After his overthrow and execution, the government protected the delta as a biosphere reserve. But there's little money to address the damage already done: one hundred thousand acres of fields and fishponds scattered throughout the area; straight channels that convey sediments to the sea instead of into the sidearms to replenish eroding wetlands. Without replenishment the delta will sink and erode, perhaps releasing a century of accumulated oil and toxins in the process.

So at the very time that nutrient pollution in the Danube was exploding, Ceauşescu was destroying the filter that might otherwise have mitigated its destructive effect on life in the Black Sea.

The other tributaries of the Black Sea were in similar shape. The Dneister—which crosses the former Soviet breadbasket—underwent similar increases in nutrients and in the 1980s was so laden with pesticides that it exceeded the USSR's health standard by six times. In addition to agricultural pollutants, the Dneiper was contaminated with fallout from the Chernobyl disaster. Damaged wetlands, dramatically increased pollution, and the alteration of the biochemical balance of the river: Together these changes have proven too much for the Black Sea to bear.

❧

IN SERGEY EISENSTEIN'S 1925 film masterpiece *Battleship Potyomkin*, the steps of Odessa look like they're a thousand feet tall. Now there's a busy multi-lane highway at the foot of the steps and beyond that a sprawling port facility built on landfill. The water's edge is now set far back from the great steps where the patriotic people of Odessa once cheered on the disgruntled sailors on the *Potyomkin*. The Bay of Odessa now seems far away, almost irrelevant to the life of this great city, which now bustles with swank shops and Mafia princes.

Out in the harbor, the Ukrainian fishing fleet is tied to its wharves, paint flaking off rusting bulkheads, nylon nets discarded in enormous heaps baking in the sun. Seagulls perch on the radar housing, piling droppings on the deck below. The six research ships of the once-prestigious Ukrainian Marine Biology Institute are moored nearby. The institute cannot even pay the salaries of its skeletal staff, so the scientists use the ships to transport commercial goods to Istanbul to earn their families' bread. The beaches along the waterfront were closed for all but one day the previous summer for reasons of public health.

In 1989 a terrible stench of rotting eggs descended on the streets of Odessa, and under the gaze of street vendors standing atop the steps, dead fish began piling up on the surface of the harbor. Out in the bay, scientists found an algae bloom of shocking proportions: Tiny phytoplankton in the water attained such dense concentrations that hydrogen sulfide gas began to rise out of the sea as dead creatures of all sorts decayed. The algae had consumed all the oxygen in the water, but submarine bacteria needed oxygen to accomplish their eternal task of decaying dead creatures. The bacteria began to strip oxygen from sulfide ions in the sea water, creating smelly, unhealthy hydrogen sulfide gas.

Creatures that weren't equipped to ferret oxygen in this manner suffocated by the millions, especially slow-swimming species that couldn't escape the vast dead zone in the bay. Light penetration to the floor fell by as much as 95 percent, killing the *Phyllophora* forests and *Zostera* sea-grass meadows, and so dooming the complex communities of living creatures that lived within them.

Monstrous blooms also appeared that summer in Burgas harbor and off the Romanian coast. Scientists at the Romanian Marine Research Institute in Constanţa recorded densities of the primary dinoflagellate at sixteen times their 1970 levels—eight billion individuals in the average liter of sea water. In the Bay of Odessa,

Ukrainian scientists recorded phytoplankton levels twenty-six times the "natural benchmark" of the late 1950s.

Only algae adapted to dim sunlight could survive below the surface layer. The *Phyllophora* forest began to disappear. It had covered ten thousand square kilometers in the 1950s and there was probably ten million tons of it. By 1992 there were only fifty square kilometers left, with a total mass of only three hundred thousand tons. All fauna native to the *Phyllophora* field disappeared: sponges, sea anemones, isopods, amphipods, shrimps, crabs, and fifty species of fish. The whole community simply collapsed.

From the brown algae prairies off Georgia to the mussel ranges off Karkinitsky Bay, the web of life was being snuffed out through suffocation and smothering. Sturgeon became endangered. Anchovy herds starved; predatory fish and dolphins vanished.

But one species in particular bucked the trend; it was in fact largely responsible for the speed and severity of this ecological apocalypse.

Soviet scientists first noticed a strange new creature in the Black Sea in 1982. *Mnemiopsis leidyi*, a bell-shaped comb jelly, was native to the east coast of North America and had ridden across the ocean to its new habitat in the ballast water of a passing freighter. The scientists had never seen one before and it took some time to positively identify the alien species; very soon it became all too familiar.

As a weakened man easily succumbs to disease, so damaged ecosystems readily fall victim to attacking forces. The Black Sea was very weak when *Mnemiopsis* arrived. Eutrophication, overfishing, the destruction of coastal habitats and wetland, and various forms of pollution had reduced the quantity, complexity, and diversity of its web of life. This diminished its ability to respond to changes or unusual events—an oil spill, an unseasonably cold winter or drought, a strange disease or invader.

Mnemiopsis had little difficulty entrenching itself. The algae blooms had been accompanied by an explosive growth in zooplankton, an umbrella term for tiny floating animals that fed on the blooms. Moon jellyfish feeding on these plentiful creatures had already shocked marine researchers with a population outburst, from perhaps 0.7 million tons in the 1960s to 500 million tons in the early

1980s. There was plenty of food for a gelatinous creature like *Mnemiopsis* and there were no predators.

Comb jellies are known for their voracious eating habits, and *Mnemiopsis* lived up to its reputation. In the 1980s it underwent one of the largest population explosions ever observed by scientists, displacing moon jellyfish and virtually every other competitor in the sea. It reproduced rapidly, sweeping the sea of zooplankton, the larvae of surviving fish, crabs, and mollusks, the only creatures that were keeping the monstrous algae blooms in check. By the late 1980s, *Mnemiopsis* had reached a total biomass of nearly one billion tons, more than ten times the weight of all the creatures landed by all the world's fishermen in a year.

Anchovy populations were already declining when *Mnemiopsis* arrived. This contributed to the comb jelly's takeover, because it had less competition from faster-swimming anchovies for the ever-growing quantity of zooplankton food. With the explosive growth of *Mnemiopsis* and the algae blooms, fish stocks suffered a sudden, devastating crash. Total landings fell to one-seventh their previous level, and twenty of twenty-six commercial fish species became commercially extinct. Anchovy landings fell by 95 percent, from 320,000 tons in 1984 to 15,000 in 1990. Dolphin and porpoise populations dropped by 90 to 95 percent. The monk seal was condemned to extinction when its last breeding refuge was blown up to make way for a Bulgarian hotel. Giant sturgeons became endangered. Oyster populations fell by 95 percent; crabs and mussels by 70 percent. Ninety-seven percent of the *Phyllophora* forests vanished, taking with them the main source of oxygen in the deeper coastal waters of the northwest shelf. Mussels once filtered tens of billions of cubic meters of sea water each day, and their departure only hastened the decline of the ecosystem.

∽

THE COSTS OF THE POLLUTION AND ECOLOGICAL COLLAPSE of the Black Sea have been enormous for the people who live around its shores.

Two million fishermen are out of work. Several people are dead and hundreds became ill from cholera and other infections after swimming in contaminated water. Beaches are closed for weeks at the height of the summer tourist season after sewage and algae blooms turned bathing into a dangerous activity; the algae blooms incubate infectious diseases by creating ideal living conditions for bacteria and even viruses. Russian scientists recorded bacterial counts ten or twenty times the normal level.

The World Bank estimates the economic cost to the region at $1 billion per year, due to losses from fisheries, tourism, and poor human health. The ecological cost cannot be measured.

☙

"THE BLACK SEA IS VERY SEVERELY DAMAGED, but it's not yet dead," Laurence Mee assured me as our boat plodded back up a Danube Delta shipping channel towards Tulcea. Dr. Mee was head of the Istanbul-based Black Sea Environment Program, an internationally funded effort to respond to the crisis, and he wanted to make one point very clear: "It can still be saved, but we must act quickly because there is very little time."

For four years, the British-born Mee had been helping to bring the coastal nations together to collaborate on rebuilding a viable Black Sea ecosystem. This was no easy task: Relations between Ukraine and Russia were tense over the future ownership of the Soviet Black Sea fleet and the Crimea and between Turkey and Bulgaria over the latter's brutal expulsion of ethnic Turks in 1989. Romania and Georgia both had long-standing grievances against the Russians. The list goes on, but Mee was able to overcome these differences by showing the scientists and new environmental ministries that they had one overriding problem in common: the Black Sea itself. The Black Sea Environment Program, with funding from the World Bank, United Nations, and European Union, has underwritten studies on the ecological catastrophe, the effects of oil spills, and the sustainable management of fisheries, and has identified the worst sources of pollution along the coast.

"It's possible to rebuild a viable ecology in the sea," Mee was saying. "The keystone species survive in small numbers scattered around the basin. If we can reduce pollution significantly to stop further damage and reduce eutrophication, we may then be able to help ecosystems recover from the bottom up. Help restore the sea grass meadows, and nature may do the rest given time. Of course it will never be like it was before—some things have been lost forever. But if the nations can act forcefully to protect the sea, there may be viable, sustainable fisheries and clean beaches once again."

Privately, other scientists connected with the Black Sea Program said it was suffering from a lack of support from foreign donors and the finance ministries of the Black Sea states for the expensive pollution reduction measures that needed to be made immediately—most pressingly the construction and rehabilitation of municipal waste water treatment plants. Mee's counterparts at the Vienna-based Danube Environment Program were encountering similar (if less severe) problems with their efforts to reduce pollution; however, nobody wanted to discuss the effect of the hydroelectric dams. In the Danube Action Plan, the destruction of the Szigetköz wetlands by the Gabčikovo dam was never mentioned, although it was making headlines in Central European newspapers every week. Slovakia didn't consider it a problem.

On the Black Sea the problem is money. In Russia and Ukraine, Romania and Bulgaria, tens of millions remain destitute, hungry, and cold, their savings wiped out by inflation, their jobs eliminated by factory closures. Homeless children compete with stray dogs for the sympathy of passers-by in Bucharest, Odessa, and Sofia. Hospitals lack supplies, staff, electricity, and heating fuel. Salaries haven't been paid for months in many parts of Bulgaria, Russia, and Ukraine. Entire cities go without electricity in Georgia, and many lie in ruins from recent civil wars. Turkey is the only Black Sea nation that isn't in the midst of a wrenching transition from socialism to capitalism. But the mountainous shores of Turkey's northern coast have been a scene of squalid destitution since the days of the Ottoman Empire and the mass expulsion of millions of Greeks, Jews, and Armenians, who were among its most prosperous and educated citizens.

To the big decision-making ministries in these countries—finance, industry and trade, energy—the ecological restoration of the Black Sea seems an unaffordable luxury. The outside world has made it clear to Eastern Europeans that they will have to take out loans to fix their environmental problems; in the calculus of the bean-counters at the finance ministries, a functioning Black Sea ecology hasn't been seen as a return on investment.

⚭

THERE IS CAUSE FOR HOPE, an optimistic scientist told me in September 1997. *Mnemiopsis*'s population appears to be collapsing. "It's eaten the sea bare. Nature is taking its course!"

Perhaps, but it appears to be a westerly course. *Mnemiopsis* scouting parties have been detected in the Sea of Marmara, the body of water between the Bosporus and Dardanelles Straits, which separate Europe from Asia and the Black Sea from the Mediterranean. This raises the specter of a comb jelly invasion of the eastern Aegean, Adriatic, and other parts of the Mediterranean already suffering ecological distress. In the early 1990s such an invasion would have been unlikely to succeed because the Mediterranean basin's living community remained diverse and reasonably resilient. As the decade drew to a close, scientists couldn't be so sure.

⚭

LET ME TELL YOU A STORY about a toxic, mutant seaweed.

Caulerpa taxifola is a delicate, bright green plant that normally grows in small, discreet clusters in tropical Pacific waters and dies if the water temperature drops below 70 degrees F. It made its way to Germany in the 1970s through the acquisitions department of Stuttgart's Wilhelmina Zoo, which displayed the plant in its tropical aquarium. Nobody is sure what happened next, but the ultraviolet light, chemicals, and human selection that followed yielded a freak form of the normally timid seaweed. "It adapts to anything—rocks, sand, and mud—to agitated currents or quiet inlets, to polluted harbors or pristine waters," says Alexandre Meinesz, a biologist at the University of Nice-Sophia Antipolis in France, who studies the weed. "Man can do nothing. Nobody can stop it."

The Wilhelmina Zoo gave samples of the pretty weed to other institutions throughout Europe, including the Monaco Oceano-graphic Museum. Museum personnel are believed to have acci-dentally released bits of the weed when they cleaned a tropical display tank. At this writing, *Caulerpa* has spread in a dense, luxuri-ant meadow across 12,500 acres of the French and Italian coasts, expanding at an exponential rate until hemmed in by water so deep and dark that even it cannot grow. This mutant strain grows to six times its normal tropical size, spreads far faster, and survives at win-tertime Mediterranean water temperatures that fall as low as 50 de-grees F. Its beautiful green meadows smother other species in its path, and new patches have appeared off Croatia, Sicily, and Majorca. Worst of all, the seaweed is toxic to most native Mediterranean animal life, which avoids it at any cost. A French re-searcher found that sea urchins resort to eating their own waste or pieces of plastic rather than touch the seaweed.

The weed spreads easily, hitching rides from one part of the Mediterranean to another by snagging on the anchors of yachts and cargo ships. A new plant can grow from a tiny clipping. Every year new colonies are discovered.

French authorities and navy divers have tried uprooting stands by hand, smothering them with salt, and sucking them away with vac-uum hoses, but the seaweed bounces back immediately. Scientists fear the weed is radically altering the entire Mediterranean ecosys-tem and are contemplating a risky solution: introducing a predatory Caribbean snail to the meadows that, hopefully, will eat *Caulerpa* in summer and die of cold each winter. Other marine experts fear this could somehow backfire, releasing a new plague of locusts on the basin.

In their desperation to bring *Mnemiopsis* under control, scientific authorities are considering something similar for the Black Sea.

Scientists generally agree that controlling the rampaging comb jelly is a necessary prerequisite to rebuilding an ecosystem capable of supporting anchovies, fishermen, and waterfront resorts. After considerable international research, scientists have thus far found only one viable strategy: An animal that feeds on *Mnemiopsis* must be introduced to the Black Sea.

Richard Harbison, a senior scientist at the Woods Hole Oceanographic Institution on Cape Cod, believes he's found the right animal. *Perprilus triacanthus*, a butterfish, is a round, six-inch fish that evolved alongside *Mnemiopsis leidyi* on the East Coast of America. It occurs in large schools and, best of all, loves to eat *Mnemiopsis* and other gelatinous animals. Because jellies are 99 percent water, *Perprilus* has to eat its own body weight in comb jellies every hour. It is also a fine food fish, so even if it is unable to conquer *Mnemiopsis*, at least it will convert many of them into a commercially useful form. If everything goes as planned.

The intentional introduction of yet another exotic species raises serious questions among ecologists, who fear the uncontrolled side effects of "monkeys playing God." Humanity has a disastrous track record in this sort of ecological engineering, particularly in the poorly understood aquatic realm. In Lake Victoria, Africa's "inland sea," native fish have been almost entirely wiped out by the Nile perch, which British colonial officials intentionally introduced in the 1960s to build a commercial-scale fishery on the lake; in the absence of native prey, the large, voracious perch started eating themselves. Scarcity and malnutrition now stalk the densely populated coastal communities surrounding the lake.

Harbison himself admits "something could go terribly wrong." But he argues that the potential risks are outweighed by the disastrous status quo. Doing nothing, he says, is the same as dumping oil into the sea and saying, "let nature clean it up."

But everyone agrees on one point: Everyone would be far better off if the Black Sea had never been allowed to deteriorate to its current state. The warning signs were there, but the sea was so large, so ancient, so seemingly invulnerable that the leaders of the surrounding countries failed to grasp the enormity of the crisis until it was too late. The sea is ringed with fishing towns that have no fish, beach resorts without usable beaches, and museums displaying ancient sculptures and artwork that pay homage to the marine life that once was synonymous with Black Sea civilization, but now is vanishingly rare.

For the rest of us, the collapse of this ancient sea serves as a warning of what is happening to marine environments around the world. The problems of the Black Sea are not so different from those confronting seas and oceans everywhere, they are just made more obvious by the sea's near total isolation.

It would be a great tragedy if that warning were lost on us.

two

Ocean Blues

Seen from space, Earth is clearly an ocean planet, a bright blue ball swaddled in clouds of water vapor. Oceans cover 71 percent of our planet's surface and clearly are its most dominant feature. By comparison, the continents seem an afterthought, an irregular brownish-green matte in which to frame our world's greatest masterpiece. For it is the oceans, not the land, that make our planet unique, setting it apart from the dozens of dead worlds in our solar system. The oceans gave our planet life, coddled and nurtured it, and allowed it to colonize the hostile environment of the land. And it is upon the profusion of life within the oceans that all of us oxygen-breathing life forms depend for our survival.

The oceans are the cradle of life, but for most of human history this fact eluded us. That's not surprising. We are a land-dwelling, air-breathing species, separated by 360 million years of evolution from the undersea home of our distant ancestors. Without special equipment we can't see properly underwater, nor can most of us visit for more than a minute or so for lack of air. Unable to explore and participate in the undersea realm, humans regarded it as a dark and threatening place, the lair of monsters and wrathful gods for whom men and ships were favored quarry. From ancient times, sheltered seas and coastal waters surrendered fish and fostered trade, but the oceans were a frightening desert, a formidable obstacle between life-sustaining continents and islands.

The Ancients knew far more about the stars and planets than they did about the undersea world, and so do we. We are better informed about the Moon and Mars than about the bottom of the ocean floor; we know more about the life cycle of stars than those of the sperm whale, giant squid, and many of the creatures sought by the world's fishing fleets.

The study of the oceans is still in its infancy. Their size and extent was not fully realized until the mid-1600s, once daring explorers circumnavigated the globe, ventured into polar seas, and ascertained the rough outlines of most of the continents. Not until the eighteenth and nineteenth centuries did mariners begin measuring the ocean's depths in a coherent way through the use of weighted lines. In the early nineteenth century Sir John Ross, a British naval commander, affixed metallic jaws to the end of a sounding line, gathered samples of deep ocean bottom ooze and was surprised to find worms and brittle stars in samples taken from more than a mile deep. The first proper oceanographic survey didn't take place until 1872, when the HMS *Challenger* circled the globe, dragging dredges along the ocean bottom to capture creatures and probe deep ocean sediments. Most oceanography is still conducted this way. Sylvia Earle, a contemporary deep sea explorer, likens the challenges of understanding the ocean bottom using such techniques to those that would confront alien visitors trying to assemble a picture of New York from net trawls of its streets and canyons. "They could amass some raw ingredients, given enough tries, but could discern little of the arrangement and no comprehension of . . . gossipy neighbors, purring kittens, parades, speeding tickets, ball games, Barney, the stock market, *New Yorker* cartoons, pizza, street gangs, or a night at the opera."

Challenger did reveal the broad physical dimensions of the ocean bottom. The continents and most other landmasses were found to be surrounded by "shelves" of relatively shallow water. These shelves sloped very gradually downward, in some places for a hundred miles or more, to a depth of a few hundred feet before suddenly dropping off as steep cliffs. The ocean bottom beyond these shelves was reasonably even and more than 2 miles deep. We now call this region the "abyssal plain." This plain was occasionally broken by enormous mountains. Some broke the surface to form volcanic islands like those of Hawaii. Others rose higher than Mt. Washington but fell short of the surface by a mile or more; these we call "seamounts." *Challenger* also discovered canyons of incredible depths, including some more than 5 miles deep.

It had long been thought that the ocean's life must be confined to its uppermost, sunlit waters. Nineteenth-century scientists reasonably assumed that life could not survive under miles of water because of cold, darkness, and, most important, crushing pressure. Water is nearly a thousand times denser than air. At a depth of ten thousand feet the weight of water above exerts a pressure of nearly 5 tons per square inch, presumably sufficient to crush any organism. To their considerable surprise, the *Challenger* crew found life throughout the abyssal plains. It was becoming clear that life existed almost everywhere in the oceans that one cared to look for it.

This realization in itself changed the way scientists thought about the distribution of life on Earth. We now know that the average depth of the oceans is about 12,500 feet (about 2 miles) and that life is present virtually throughout. Compare that to the habitable volume of the land. Fertile soil extends, on average, only a few feet below the surface, while forests reach an average of a hundred feet into the sky. Such analysis shows that the oceans comprise 99.5 percent of the Earth's habitable space. Humans, accustomed to considering themselves at the center of things, turned out to be living on the living planet's periphery.

After World War II a tiny but wonderfully rich fraction of this vast underwater living space was made accessible to ordinary people for the first time. Working in German-occupied France during the war, Jacques Cousteau and Frederic Dumas perfected the aqualung, a self-contained underwater breathing apparatus (SCUBA). This reliable and inexpensive technology has since allowed millions of people to see and explore the ocean on its own terms. Cousteau and those who followed him beneath the waves found forests, meadows, mountains, gorges, and coral reefs inhabited by an incredible profusion of life. Cousteau's films allowed millions more to see soaring flocks of fish, herds of mammals, the birthing and nursing of whales, and ocean bottoms covered in plants and animals of every conceivable shape and color. Rather than a desert, the sunny shallows of the continental shelves were revealed to be tranquil, beautiful, and teeming with life.

Such is the strength of our bias towards the land that not until we started searching for signs of life on other worlds did we really begin to understand the significance of the oceans to life on Earth. Venus and Mars, our nearest and most similar neighbors, are arid, oceanless, and surrounded by atmospheres consisting almost entirely of carbon dioxide (CO_2). Earth, unique in the known universe, has an atmosphere dominated by nitrogen and oxygen and a surface dominated by liquid water. It also has life. Scientists are reasonably certain that these three traits are interrelated.

When life formed on Earth more than 3 billion years ago the atmosphere was probably much like that of Venus today: mostly CO_2. Conditions on land were not conducive to life. The surface would have been buffeted by volcanic eruptions, frequent meteor collisions, and intense radiation. But the watery brine of the oceans offered protection and a relatively stable environment; it also contained the chemical ingredients of life in the perfect medium, liquid water. Liquid water is widely considered a prerequisite for life since it is necessary for both photosynthesis and the biological activation of DNA and RNA, the carbon-based molecules that store an organism's reproductive information. Furthermore, large quantities of free oxygen (O_2) can probably only exist in the presence of plants. Because of photosynthesis, plants absorb CO_2 and release oxygen. Without a constant, massive influx from plants, any supplies of free oxygen in the atmosphere would be destroyed by solar ultraviolet radiation. Life evolved in the oceans and over many millions of years its byproducts altered the chemistry of the atmosphere sufficiently to allow the evolution of oxygen-breathing land animals like ourselves.

After 365 million years ashore, we are only beginning to explore the watery kingdom from which all life sprang. Sadly, those explorations are revealing that our power and ignorance are rapidly destroying the living oceans. The cradle of life is becoming a watery grave.

❧

I WRITE THIS BOOK AT THE CLOSE OF ONE MILLENNIUM and the beginning of a new one. Doomsayers, religious zealots, and computer bug alarmists warn that the end of civilization is at hand. When the clocks roll over to 2000, nuclear war, electronic pestilence, and a plague of extraterrestrial visitors are to presage the Second Coming and the end of the world. But if you're reading this, then civilization must have survived these experiences. Perhaps we've gained some insights, or at least sobered up sufficiently, to begin to tackle the apocalyptic challenges that remain.

The greatest challenges facing humanity in this new century are environmental. Life on Earth stands at a frightening crossroads. There are getting to be too many of us. Six billion people live on Earth today, twice as many as in 1960. Thankfully, the rate of growth has slowed, but every year global population still increases by 80 million. World population is likely to stabilize at a staggering 10 ½ billion by the end of the twenty-first century. Ninety-eight percent of that growth will be in developing countries, most of which already are unable to provide jobs and basic services to their exploding masses. We residents of wealthy countries may continue to live comfortable lives, but we will be continually confronted by the human misery, wars, famine, and instability that surround us. We will also face an unparalleled environmental crisis as our sheer numbers overwhelm the complex natural systems that sustain life in what appears to be an otherwise lifeless solar system.

We live in the midst of a mass extinction, the greatest extermination of living species since the sudden end of the dinosaur age 65 million years ago. Every year, between seventeen thousand and one hundred thousand species vanish from our planet as their habitat is cleared or non-indigenous species are introduced through human commerce; the wide range is due to our ignorance of how many species, particularly insects, live in threatened rain forests. Stuart Pimm, a biologist at the University of Tennessee, compared the extinction rate over the last century with the background rate calculated from the fossil record. He concluded that extinction rates have increased by between one hundred- and one thousand-fold during the twentieth century. Paleontologist Richard Leakey estimates that 50 percent of the Earth's species will vanish during the twenty-first century. Humankind, Leakey maintains, now uses 40 percent of the energy available to sustain life on Earth, leaving little to power the ecosystems that maintain the planet's atmosphere, climate, and other life support functions. That percentage will increase with population growth, imperiling not just other creatures but ourselves.

Until the 1990s people generally assumed that this wanton destruction of nature stops at the ocean's shore. I was told from elementary school on that the oceans are our strategic reserve of food and resources; they are simply too vast to be seriously affected by human activity. If we run out of arable land to feed the growing masses then we'll learn to farm the seafloor. Not enough meat protein to go around? Not to worry, we can always draw on the ocean's limitless supply of fish. Not enough living space? Too much ultraviolet light coming through the ozone hole? Perhaps we can build underwater cities. We can trust the oceans to fill the gap, my mid-1970s reading curriculum suggested, until we colonize the Moon and Mars. Perhaps then we could simply abandon our damaged planet like some blighted inner city neighborhood.

In reality the oceans are not in much better shape than the land. During the second half of the twentieth century, life in the oceans was decimated by the combined effects of human activities. From frigid polar seas to the sweltering tropics, fish populations have collapsed at the hands of a potent alliance of greed, politics, and technology. Entire seas have been ravaged, their living bounty wasted by pollution and the introduction of non-indigenous species. Lifeless zones have formed and spread in bays and coastal waters, places from which animals must evacuate or suffocate for lack of oxygen. Marshes, sea-grass meadows, and mangrove forests—the three great nurseries of the sea—are being lost at a staggering rate as coastal land is developed. A warming climate and an expanding ozone hole are stressing already weakened ecosystems, causing mass die-offs of coral reefs in the tropics and declines of krill and penguins in the Antarctic. In a third of the Pacific, warmer water temperatures appear to have triggered a 70 percent decline in zooplankton, the "fish food" on which all higher life depends. These assaults interact and reinforce one another.

People already feel the consequences as damaged marine systems stop delivering myriad "free" services. Islands wash away during storms. Formerly stable shorelines sink and erode. Tropical nations watch tourism and fisheries decline with the coral reefs that sustained them. The quality and quantity of seafood is diminishing, imperiling over one billion people who depend on it for their daily protein needs. Those who live close to the sea are hardest hit. Fishermen from Atlantic Canada and New England to Ireland and the Philippines find they can no longer make a living from the sea. Residents of low-lying areas and small island nations face the specter of permanent inundation from sea-level rise.

This is a book about the deterioration of life in the oceans and along its shores. It is a story much like that of the destruction of tropical rainforests, only far larger in scale. A vital and irreplaceable part of the living planet is being destroyed before we even begin to understand it. There is plenty of blame to go around. Politicians, developers, corporations, tourists, you and I all play a role. But the underlying problem is ignorance, a failure to comprehend that despite their scale, the oceans are as finite and destructible as the forests, jungles, rivers, and lakes of the continents. It's not too late to mend our ways. The oceans can and should be protected, and the effort would make our global economy more efficient, less wasteful, and better able to provide for a world of 10 billion. First, however, we must understand what the oceans are and how we are causing them so much trouble.

∞

TO UNDERSTAND WHAT IS HAPPENING TO THE OCEANS, there are a few things one must first know about life within them.

As on land, most marine life is sustained, directly or indirectly, by energy from the sun. Plants harness the energy in sunlight to build organic matter through photosynthesis. Other creatures eat the plants to access this energy and, in turn, are eaten by predators. Plants, consumers, and predators eventually die and, if not consumed first, their bodies are broken down into organic matter by bacteria, completing the cycle.

In the sea, photosynthetic production is dominated by microscopic algae suspended in enormous numbers within every drop of sea water. Except for a few shallow water plants like seaweed, kelp, or sea lettuce, virtually all the food production in the oceans originates with these tiny, often single-celled plants. Collectively they are referred to as "phytoplankton," from the Greek words *phyton* (plant) and *planktos* (wandering). These phytoplankton are the meadows of the sea, the vast range lands upon which the ocean's herbivores graze. There are thousands upon thousands of species, but the diatom family is by far the most numerous and best understood. Diatoms are single-celled plants encased in tiny glass shells of their own making. They reproduce by dividing in half, shell and all, and can bloom in such enormous quantities that they change the ocean's color from blue to green. In addition to providing food for higher animals, diatoms may also play a critical role in the regulation of the Earth's atmosphere.

Diatoms are also exquisitely beautiful. I had my first glimpse into the diatom's world aboard a pitching ship in the midst of Antarctica's Southern Ocean. The scientists had brought millions of these creatures up from several hundred feet of water in sample bottles that morning and filtered them into a sort of diatom concentrate to better examine them. Patrick Neale, a phytoplankton expert at the Smithsonian Institution, made me a slide and left me with the microscope. When he returned an hour later I was still fixed in place in the cramped, heaving cabin, eyes glued to this window into a newfound universe. On this single slide were more diatoms than I could possibly count. Some were shaped like delicate pillboxes and elongated chains, some had long glass spikes growing from their sides, others resembled segmented worms, firecrackers, spring peas, and necklaces with tablet-shaped beads. Each movement of the slide brought new and unexpected creatures into view. Under special light their chloroplasts could be made to glow blood red or purple. It reminded me of astronomers' pictures of stars, gas clouds, and spiral galaxies—so much so that at times I forgot that I was on a ship at the bottom of the world, imagining instead that I was peering into the cosmos through a powerful telescope. Every bucket of the sea contains an entire galaxy of life.

Life in the ocean is not distributed equally. It tends to concentrate near areas where phytoplankton are found in great numbers. Phytoplankton productivity is constrained by the availability of two things: sunlight and nutrients. These are not always found together. Below two hundred or three hundred feet light doesn't penetrate well enough for the tiny algae to thrive, so virtually all of the primary production in the oceans takes place near the surface. They also need nutrients—phosphorus and nitrogen—but these tend to sink to the dark bottom sediments. However, some parts of the world ocean retain unusually high levels of these critical elements near the surface and thus support huge phytoplankton blooms. These nutritive surface waters include coastal areas that receive runoff from the land and "upwelling" areas where deep ocean currents rise to the surface laden with sunken nutrients. Not coincidentally, these areas are also the world's most productive fishing grounds.

Where there is phytoplankton, there is life. Swarms of tiny animals feed directly on individual diatoms and dinoflagelletes (another type of microscopic algae). Some of these animals are single-celled protozoans that absorb the plant cell directly into their own. Most are seed-sized crustaceans, which scoop the little algae into their mouths or eat their cousins who do so. Under a microscope they look like tiny shrimp or lobsters, with many legs and antennae protruding from a segmented shell. They fall into an alphabet soup of scientific families: copepods and isopods, decapods and amphipods. These near-invisible creatures are the main fish food for herring, mackerel, anchovies, and countless other plankton-eating creatures. A single herring stomach has been found to contain sixty thousand copepods. Considering that we once caught two million tons of herring every year—and many times as many herring-eating fish—the number of copepods required to support our fisheries bends the mind.

Big fish eat small fish, of course. Herring and other plankton-eating fish are gobbled up by predators like cod or bluefish which, in turn, are eaten by tuna, sharks, seals, and people. But many extremely large animals feed directly on copepods and other animal plankton. They are the primary food of the great baleen whales, including the right and finback, which strain the sea with their sieve-like baleen, catching speck-sized animals by the tons. The world's largest shark, the docile forty-foot whale shark, also feeds directly on plankton.

The creatures of the deep ocean bottom also depend on the plankton—both animal and plant. As diatoms, copepods, and other creatures die they slowly sink to the bottom in an unending snow. Most get eaten before they fall into the deep, but the remaining detritus provides food for bottom-dwelling scavengers and worms. The feast at the surface leaves ample scraps for uninvited guests below. Because food is limited, deep-dwellers are long-lived and slow to reproduce, which makes them unusually vulnerable to fishing pressure.

Within this broad scheme, life in the ocean has many tricks and adaptations, some of them quite surprising. Tropical waters are nutrient-poor since their relatively uniform warmth discourages upwelling. Yet they are home to coral reefs, the most diverse and productive ecosystems in the world ocean. Corals, as discussed later in this book, found a way around the problem by incorporating plants and animals into the same colonial organism, cycling energy and nutrients far more efficiently than is otherwise possible. Corals have also built elaborate structures around which thousands of specialized animals evolved, cycling nutrients so efficiently among themselves that the reef ecosystem teems with life in the midst of what ought to be a biological desert.

Some parts of the ocean are more critical than others. Most higher life in the oceans relies on near-shore shallows for at least part of its life cycle. Salt marshes, mangrove forests, sea-grass meadows, and kelp forests are often called the "nurseries of the sea" since so many fish, shrimp, lobsters, crabs, and other creatures start their lives in these protective, food-rich environments. Without these nurseries, large numbers of marine organisms are doomed, from Gulf shrimp to Florida manatees. They represent the ocean's Achilles' heel, for they are located in coastal areas, where they become casualties of human enterprise. Other species spawn and develop at particular offshore locations uniquely suited to their purposes. Some bottom-dwelling fish require a particular type of gravel bed located near a food-rich upwelling area and only a few dozen locations will do. Antarctic krill hatch in dark abyssal depths around Antarctica, but the larvae then rise in search of sea ice, under which they find protection and food in the form of a remarkable algae that grows on the underside of the ice. Less ice means less krill, the staple diet of Antarctica's penguins, seals, and whales.

∞

WE ARE JUST BEGINNING TO EXPLORE the marine realm and unravel its secrets. Marine scientists have learned to expect the unexpected. Their discoveries in the ocean depths have shocked the scientific community, turning basic assumptions upside down, often raising more questions than answers.

The fossil record shows that many unusual creatures once lived in the world's seas. There were four-foot-wide clams that housed schools of a hundred or more small fish inside their shells, perhaps protecting them from twenty-foot-long bony fish, whose bulldog-like heads were filled with pointy teeth. There were fast-moving fish-lizards with dolphin-like bodies and pointed snouts. Vast schools of armored squid were stalked by 10,000-pound plesiosaurs, a long-necked sea dragon resembling the Loch Ness Monster. Like many of these creatures, the coelacanth, a five-foot-long fish with tasseled fins, was thought to have become extinct about 80 million years ago. This primitive animal interested evolutionary biologists because it is a surviving record of the critical period in which fish prepared to crawl out of water and onto dry land. It had four flipper-like fins stunted with the beginnings of legs and was closely related to freshwater lungfish, whose respiratory system allows limited air breathing. But unlike lungfish, coelacanths were wiped out along with the dinosaurs.

Or so we thought.

In 1938 South African fishermen netted a flesh-and-blood coelacanth while trawling in the Indian Ocean. It was a large and beautiful fish, weighing over 100 pounds and covered in steel-blue scales with pale spots. This astounding discovery prompted further quests to find other "living fossils" and reportedly inspired the classic horror film *The Creature from the Black Lagoon*. Coelacanths eluded further capture for many years but were eventually found to number less than a thousand, all apparently living in deep underwater caves in the Comoros Islands. They were thought to be a fluke, a remnant population that had survived the last mass extinction by hiding out in caves. But in 1997 a honeymooning biologist found one in a fish cart 6,000 miles away on the island of Sulawesi, in Indonesia. Coelacanths were more widespread than anyone had suspected. The biologist, Cincinnati-born Mark Erdmann, returned to Sulawesi in 1998 and received a live coelacanth from a local fisherman. Erdmann was filmed swimming with it in a lagoon before it died.

Unfortunately a great many coelacanths are dying at the hands of scientists, so many in fact that they may be driven to extinction. Aquariums and museums all want coelacanth specimens, and hundreds of the rare fish have been killed. Many more have reportedly been slaughtered to feed a perverse Japanese market for this rare fish. Some wealthy Japanese apparently believe that a creature able to cheat extinction for so long must contain the secret of everlasting life. Dealers reputedly pay Comoros fishermen to capture coelacanths, whose spinal fluid is extracted and sold for many thousands of dollars per teaspoon.

If they can be studied without being exterminated, coelacanths may tell us a great deal about how animals moved from sea to land. But a more recent discovery may explain how and where life on Earth actually began.

Most of us were taught that all life on Earth is sustained, directly or indirectly, by energy from the sun. Plants capture solar energy through photosynthesis, grow, and are in turn eaten by animals that are then consumed by other animals. Life on Earth is solar powered.

We've since learned that that isn't entirely the case. While exploring volcanic activity on the abyssal ocean plain near the Galapagos Islands, researchers aboard the submarine *Alvin* found the last thing they had expected: an exuberant ecosystem thriving on raw chemicals pouring forth from the bowels of the planet.

Under a mile and a half of water, *Alvin* found hot springs and vents that emit a flow of mineral-rich, superheated water. The minerals collect to form tall chimneys around the vents, strange structures that can reach the height of a fifteen-story building, upon which gather oases of exotic creatures. There are red shrimp, blind white crabs, diminutive lobsters, shoals of mussels, and enormous clams. Large fish swim amid sea anemones and thickets of tube-shaped worms anchored to the seafloor. It is surprising enough that all these creatures can survive under enormous pressure and total darkness, in waters that turned out to be not only acidic but heated to 600- and 700-degree F temperatures. But what really shocked the scientific community was how the ecosystem acquires its energy. The vents are teeming with simple, bacteria-like microbes that grow by consuming sulfurous compounds emitted by the vents. This oasis of deep-sea life is powered not by photosynthesis but by chemosynthesis. Not only that, but later research showed that the microbes are among the most primitive life ever found and more closely related to the earliest forms of life than any other. The most recently discovered category of life turns out to have been around longer than all the rest.

These organisms may have been the first on Earth. There is growing evidence to support the theory that life began with organisms like these. Experiments have shown that such hot vents (many more have since been discovered) contain all the necessary chemical building blocks of life and that the superheated conditions facilitate the appropriate reactions between them. Given the incredibly hostile conditions on the Earth's surface at the time, the ocean bottom may have been a more satisfactory place for life to have gained a foothold. How ironic that life may have begun in an inky abyss, acidic and searingly hot, a byproduct of the radioactive decay of heavy metals in the inferno of the planet's core.

What other secrets are encoded in the tissues of the ocean's many creatures? How many clues will be erased before we find them?

∞

OUR ASSAULT ON THE OCEANS IS MULTI-PRONGED and implicates just about everyone. From fishermen and bureaucrats to farmers and factory owners, from real estate developers and dam builders to owners of ships, golf courses, and powerboats, the burden is broadly shared and so must be the solutions.

Of the many symptoms, the collapse of the world fish populations has received the greatest attention. The statistics collected by the United Nations Food and Agriculture Organization (FAO) are daunting. Of the world's fifteen major fishing regions, eleven are in decline. Overall catches have dropped by more than 50 percent in the Southeast Atlantic and 40 percent off the American and Canadian Atlantic coasts. Sixty percent of the two hundred major commercial fish species are "fully exploited" or in decline. Many high-value fish species have collapsed: salmon in the Pacific Northwest, bluefin tuna in the Atlantic, and Nassau groupers in the Caribbean, to name a few. This has forced fishermen to move to lower-value fish like Alaska pollock and horse mackerel, previously spurned as neither tasty nor useful. As they fish lower on the food chain, fishermen compound the problem by depleting the food supplies available to larger fish, thus reducing their chance of recovery. Small, low-value schooling fish now dominate world fishery landings. In 1995 six such species accounted for 25 percent of total fish landings worldwide. More valuable fish are increasingly difficult to find.

How could this happen? The main reason is the development of new technologies that allowed fishermen to scoop up the ocean's fish faster than they could reproduce. Refrigeration, factory automation, and shipbuilding innovations allowed the construction of huge factory ships able to process and freeze their catches on board and thus operate many thousands of miles from port in all but the worst weather. First deployed against the great whales, factory vessels drove themselves out of business in a few decades by exterminating their prey. The largest animal ever known, the blue whale, was reduced from an estimated two hundred thousand individuals in 1900 to no more than two thousand today. Right whales fell 98.5 percent to three thousand, humpbacks by 92 percent to ten thousand, Sei by 88 percent to an estimated twenty-five thousand. The bodies of these magnificent creatures were converted into fats, oil, and meat byproducts to produce such necessities as margarine, canned pet food, tennis racket strings, perfume, and cosmetics.

Once whaling ceased to be commercially viable, factory ship designs were modified to hunt fish, which they now destroy with equal efficiency. Radar, fish finders, and, later, satellite navigation allow the enormous ships to locate and capture fish schools with lethal accuracy. Today's factory-freezer trawlers haul nets large enough to swallow a formation of twelve Boeing 747 jumbo jets. Unloaded and resupplied by oceangoing tenders, these ships can fish twenty-four hours a day, seven days a week, for many months on end in any part of the world ocean. They are so efficient at stripping the sea of fish that many ships can no longer find enough fish to remain profitable. These vessels played a central role in the destruction of the once-great fisheries off New England and Atlantic Canada. The factory fleets then moved on to decimate Greenland turbot, North Pacific ocean perch, and Bering Sea pollock. At this writing they appear to be doing the same to the pollock fishery off Alaska and are implicated in the starvation of stellar sea lions, which eat pollock. The fish are converted into imitation crabmeat, fishmeal, and fast-food fish fillets.

Nor is the damage limited to the targeted fish species. Fishermen capture huge quantities of marine life they don't want, since they are either of the wrong species or size. FAO estimates that for every three tons of fish landed at the dock, another ton of unwanted creatures is thrown overboard dead or dying. That works out to 27 million tons of so-called bycatch killed every year. The annual death toll includes eighty thousand albatross drowned after taking bait on floating hooks trailed on miles of line behind "longliner" fishing vessels; one hundred thousand sea turtles ensnared in shrimp trawls; hundreds of thousands of seabirds ensnared in enormous open-ocean driftnets; so many thousands of dolphins and porpoises drowned in traps and gill nets as to decimate seven stocks and threaten forty-six others; untold billions of juvenile cod, redfish, pollock, and other valuable fish; and countless sharks, squids, crab, starfish, sponges, anemones, and other creatures.

Some fisheries are more damaging than others. Because their nets must be fine enough to catch tiny prey, shrimp trawls have enormous bycatch. Eight to nine pounds of unwanted creatures die for every pound of shrimp caught in the Gulf of Mexico. When fishing trawlers work in tropical waters bycatch can also reach 80 to 90 percent of the total haul. Off Hawaii, tuna and swordfish boats cut the fins off tens of thousands of sharks and toss the carcasses back overboard. The practice was only recently banned on the U.S. Atlantic seaboard. Top grade fins sell for more than $100 per pound in Asian markets and are used to make soup.

A particularly disturbing development is the advent of "biomass fishing." Huge trawls with extremely fine mesh gather anything larger than an American quarter. Most of the net's contents are ground into meal and fed to farmed shrimp, fish, and poultry or simply spread on fields. The ecological effects can only be guessed at.

∞

ONCE DAMAGED, THE RECOVERY OF FISH POPULATIONS is often hampered by less direct stresses; in some cases collapses are caused by them. Most scientists agree that the most serious of these is the destruction of the physical habitats suitable for various species to hatch, grow, feed, and breed.

Recall that many species rely on salt marshes, sea-grass meadows, mangroves, or kelp forests for survival during at least part of their life cycle. Because they also tend to occupy potentially valuable coastal real estate, a large proportion of these "nursery" habitats has been destroyed to make way for real estate developments, golf courses, marinas, and shrimp farms. Worldwide development and pollution are estimated to have already destroyed or degraded half of all coastal wetlands since 1900. America has lost more than half the marshes, mangroves, and sea grass the country started with. The Philippines has lost 70 percent of its mangrove forests since the 1920s. The World Resource Institute estimates that just the loss of mangrove forests—about half the world total—translates to a loss of 4.7 million tons of commercial fish. The clearing of mangroves is killing reefs and eroding islands in Belize, threatening the liveli-hoods of entire communities. The Black Sea's sea-grass meadows and kelp forests were smothered by sewage and fertilizer run-off from the land. In the Mississippi River Valley the construction of flood control levees and gas-field access canals is causing the south-ern quarter of Louisiana to sink away, including the bayous that not only nourish much of the Gulf of Mexico's marine life but protect New Orleans and other communities from hurricanes.

Coral reefs, the foundations on which most tropical marine life rest, are declining at a staggering rate. Clive Wilkinson, a coral reef expert at the Australian Institute of Marine Science, estimates that about 10 percent of the world's coral reefs are already dead or dying and another 30 percent are expected to decline significantly over the next fifteen years. People are killing reefs in a variety of ways. Cutting down mangroves is particularly destructive as it not only wipes out many reef creatures but also triggers serious erosion, which smothers the living corals in plumes of soil. Overfishing by traditional fishermen has become widespread as tourists, wholesale marketers, and population growth create incentives to catch ever-larger numbers of ever-smaller fish. In many countries this process has reached an illogical extreme. In the Philippines, Micronesia, and Jamaica, people blast reefs with dynamite and other explosives to stun and kill all marine life over a wide area. The floating creatures they gather represent a one-time bonanza. The fishermen kill the goose that lays their golden eggs. To supply live fish to elite Asian restaurants and the American home aquarium market, village divers employ cyanide spray bottles to stun tropical fish; the cyanide kills corals over wide areas. But the biggest threat to corals comes from climate change and ozone loss, which create conditions for a deadly syndrome known as "coral bleaching."

Nor are ocean floor habitats safe from human mischief. The floor of the continental shelves—where most ecological activity takes place—is covered with ledges, boulders, cobbles, and a variety of plants and other creatures that provide protection and living habitat for other bottom-dwelling creatures. Trawl nets and other gear dragged across the bottom by fishermen destroy these structures like gigantic plows. Les Watling of the University of Maine has hours of "before-and-after" videotape showing the effect of such trawls: gardens of life in one segment, mud and debris in the other. Watling and other scientists studying the problem compare the ecological disruptions to the clear-cutting of forests on land.

In vast near-coastal areas and in semi-enclosed seas the water itself has been rendered sterile. The problem is nutrient pollution, the smothering deluge of sewage, manure, and chemical fertilizers from land-based activities. Since World War II, rising populations and the increasingly intensive agriculture and livestock operations needed to feed them have caused an explosion in nutrient run-off. While some nutrients are good, too large a quantity spells disaster. Phytoplankton productivity is limited by the availability of nutrients in sea water, and where there are excessive levels these microscopic algae explode in such massive blooms that grazers can't keep up. The dead algae fall to the bottom to be decomposed by bacteria, a process that consumes large amounts of oxygen—so much that often little or none is left to sustain anything else. This condition is called hypoxia. When hypoxic conditions occur, all animal life that cannot swim away suffocates. This is how the Black Sea's shallow, life-bearing shelves were laid waste, setting the stage for the ecological collapse of the entire basin. Hypoxia has also become a chronic problem in the Gulf of Mexico, where a seven thousand square-mile "Dead Zone" appears off the Louisiana and Texas coasts every spring and summer, disrupting shrimp and fish migrations and wiping out bottom fauna. Seasonal hypoxia afflicts large portions of America's Chesapeake Bay and New York Bight, the Adriatic, North, and Baltic Seas, the Inland Sea of Japan, the Yellow Sea, the Persian Gulf, and bays and harbors the world over.

Of course, this is not the only sort of pollution that occurs. Producers of hazardous substances have always looked to the oceans as a convenient dumping ground. Toxic effluents, radioactive waste, and other poisons seem to disappear under the waves, but they are most definitely heard from again. Rachel Carson's *Silent Spring*, the 1962 book that in large part triggered the birth of the green movement, showed how synthetic substances may become highly concentrated as they work their way up the food chain. In the marine realm, too, trace contaminants concentrate in the bodies of organisms, sometimes killing them directly but more often reaching lethal levels in the organs of predators.

Tributylin, the organic form of tin, is used in most marine paints since it prevents the growth of barnacles and seaweed on the underside of ships. It dissolves into the water in trace amounts but becomes concentrated in the bodies of shellfish and other creatures that feed by filtering huge quantities of sea water through their bodies. Tributylin is thought responsible for dramatic reductions in marine snails in many harbors worldwide. After a mass die-off of California sea otters, scientists found heavy concentrations of the substance in their livers. Researchers in the Canadian Arctic have found DDT, PCBs, heavy metals, and other industrial chemicals at every level of the food web. They have reached potentially harmful levels in the bodies of polar bears, seals, walruses, and whales, and in the milk of Inuit women. Many Inuits—the indigenous people of the Arctic—have concentrations of certain pesticides in their bodies that exceeded safe levels by twenty times. Chapter 6 describes how radioactive products of U.S. atomic and hydrogen bomb tests caused birth defects, cancers, and numerous deaths among Marshall Island residents who were exposed after eating contaminated fish, crabs, and other organisms.

❧

NOT ALL OF THE ECOLOGICAL DAMAGE comes from harming marine life directly. A great deal of havoc can be caused simply by moving it around. The most common agents of what scientists call "invasive species transfer" are oceangoing tankers and container ships. When light on cargo, most ships are obliged to pump water into their holds to maintain their seaworthiness. This ballast water contains dozens of plants and animals, some as adults, but most in the form of eggs, larvae, or juveniles. On reaching its destination halfway around the world, a ship will discharge some or all of its ballast water and, in the process, introduce huge numbers of alien species to the surrounding environment. Worldwide, the National Research Council estimates that three thousand species are picked up in ballast water every day. Many of these species don't survive either the trip or the conditions of their new environment. Some do and become established in their new circumstances. Lacking natural predators or having overwhelming advantages over their prey, some intruders completely take over, exterminating competitors and turning the ecosystem upside down.

"Once an exotic species is established, trying to remove it is like trying to put the toothpaste back in the tube," James T. Carlton, an invasive species expert at Williams College in Massachusetts, told me. Carlton says he and his scientific colleagues are finding that many marine ecosystems in the United States and around the world are already dominated by creatures introduced over the past century, making proper damage assessments impossible. "The world," he says, "is just itching with invasions."

At age five, when I first started exploring tidal pools on the Maine coast, I found small green crabs in just about every hiding place I explored. These ubiquitous little shore crabs with their tickling bite and amusing sideways gait were, to my mind, emblematic of life in Maine's cold blue ocean. Only a few years ago I was surprised to learn that this species—the *European* green crab—is a newcomer to North American shores, and a disruptive one at that. It was introduced to America's eastern seaboard in the early nineteenth century, first appearing only between New Jersey and Cape Cod, but later spreading as far north as Nova Scotia and south to Maryland. The green crab is a voracious eater, preying on oysters, clams, mussels, other crabs, and small fish with such effectiveness that major ecological changes ensue. It has recently established itself in Japan, South Africa, and the Pacific Coast of America, where it is causing considerable concern among fishermen and ecologists. Its larvae probably hitch rides on ships. First reported in San Francisco Bay in 1989, green crabs have spread as far north as Washington state, displacing native crabs eaten by people. On the East Coast it was implicated in the collapse of the Atlantic softshell clam industry in the 1950s and in reductions of oyster harvests. The green crab is not a good substitute for fishermen since its small size makes it difficult for people to eat or process.

As ships get larger and more numerous, such invasions are becoming increasingly commonplace. San Francisco Bay, a busy shipping port, is home to at least 212 exotic species. The fish population is now a bizarre mix of Mississippi catfish, East Asian gobies, Japanese carp, and aquarium goldfish. The bottom is controlled by Chinese mitten crabs (which can harbor human parasites and whose burrowing causes levees to collapse) and Asian clams (which filter out virtually all plankton, starving out native fish). A new species takes hold in the Bay every twelve weeks on average. Exotic invaders tend to wreak the most havoc in ecosystems already damaged by other stresses. A North American bristle worm now dominates the bottom of Poland's highly polluted Vistula lagoon. *Mnemiopsis leidyi* snuffed out most other life in the Black Sea after massive algae blooms, overfishing, and pollution weakened native communities.

∞

THE MOST GENERALIZED THREAT TO LIFE in the oceans also has the most potential to trigger widespread disruptions of human economies, societies, and general well-being. Our release into the atmosphere of substances that alter climate and increase ultraviolet radiation at the Earth's surface has not spared the oceans. Both are implicated in ecological disruptions reported throughout the world ocean. Climate change—whether it is caused by humans or not—

has the potential to disrupt the entire scheme of currents and nutrient flows on which most marine life now depends. It is also expected to put cities, provinces, even entire countries under water.

Ultraviolet (UV) radiation is extremely harmful to living organisms. Many scientists believe that life on land could have taken hold only after the formation of the stratospheric ozone layer, which protects the Earth's surface by absorbing the most damaging UV rays. Unfortunately, the ozone layer has been damaged by our manufacture and release of chlorofluorocarbons (CFCs) and other ozone-depleting substances. This has resulted in the formation of an enormous ozone hole over Antarctica each spring, beneath which organisms are bathed in triple the normal levels of high-energy UV-B. But ozone depletion effects more than just the Antarctic. Scientists have recorded a second ozone hole over the North Pole and reduced ozone levels throughout mid-latitudes of both hemispheres. Increased UV is already implicated in widespread damage to coral reefs through a life-threatening syndrome called "coral bleaching." Increased UV is also known to cause population reductions in some juvenile fish, such as the northern anchovy, and may contribute to the failure of other species to recover from population collapses caused by overfishing. Although the production of CFCs has been curtailed under a 1979 international agreement, ozone depletion will continue to worsen in coming decades as existing stockpiles are used up and previously released CFCs complete their slow migration into the stratosphere.

The planet is also heating up. Mean surface temperatures increased by 0.3 to 0.6 degrees C (0.5 to 1.0 degrees F) during the twentieth century. As of this writing, all fourteen of the warmest years since 1860 have occurred in the last two decades, with 1998 the hottest year in recorded history. Of these facts there is little doubt. What most scientists and politicians debate is whether the observed warming is due to increased levels of carbon dioxide and other heat-trapping gases in the atmosphere—gases released by industry, automobiles, and power plants. The Earth's climate is an incredibly complicated system that we are only beginning to understand, so scientists can't be 100 percent certain of either the causes or the effects of global warming. But based on the balance of the evidence, the main international assessors of climate change—a large and conservative United Nations body called the International Panel on Climate Change (IPCC)—believes there is "a discernible human influence on global climate."

Climate change affects marine life in several ways. The first is perhaps the most obvious: A warmer Earth means warmer oceans. Water temperature is one of the key factors influencing where a particular marine organism can live, feed, and reproduce. The warming of ocean surface layers—where most marine life lives—will stress some populations and probably wipe out others that already live at the warm extreme of their tolerance range. These changes can be very dramatic. In the northwestern Pacific, the enormous swath of ocean bordering California and southern Alaska, average surface temperatures jumped by two degrees F in 1977 and have remained at these levels ever since. John McGowan of the Scripps Institution of Oceanography found that over that period there has been a 70 percent decline in zooplankton, the tiny animals that, directly or indirectly, feed virtually all higher life. McGowan has documented an ecological crisis extending across a third of the North Pacific, from the Gulf of Alaska to the southern California coast. A once-common seabird, the sooty shearwater, declined by 90 percent. Most fish populations have fallen by 5 percent per year since 1986, and near-shore species like kelp, urchins, and abalone have collapsed while warmer water species have moved in. The short-beaked dolphin, a species that prefers warm water, has undergone a twenty-five-fold increase in the waters off California, while Alaska's cold-water fur seals and sea lions have been dwindling. Nobody is certain if the increased water temperatures are due to global warming, but they foreshadow the disruptions we can expect as the oceans grow warmer.

Unusually warm water temperatures are stressing coral reefs worldwide and are believed to be the main cause of coral bleaching. During 1998, tropical sea surface temperatures were the highest on modern record, topping a fifty-year trend in some areas. The warmer temperatures contributed to the most severe coral bleaching events ever recorded, in many regions affecting 75 percent of corals. Widespread bleaching was reported in all coral reef regions worldwide except the Central Pacific, and affected corals in as much as 150 feet of depth, rather than the 45-foot depth limit usually associated with bleaching events.

The Antarctic is also experiencing ecological crises brought on by warming sea temperatures. The Southern Ocean's extraordinarily rich marine ecosystem relies on millions of square miles of sea ice that form around Antarctica every winter. Diatoms and other algae grow in enormous quantities on the underside of the ice, providing the main food supply for krill, which in turn feed almost everything else. But warmer seas have meant that less winter sea ice forms in the northerly Antarctic Peninsula region each year. Reduced sea ice is implicated in dramatic changes in the abundance not only of krill but of various types of penguins and seals. So far, warming has been confined to the Antarctic Peninsula, which has a milder climate than the rest of the continent. If the climate continues to warm as the IPCC predicts, the rest of the Antarctic may be affected, with disastrous results for the creatures that live and feed there.

Global warming has a second set of effects. The world's sea level is rising, placing both human and marine near-shore life in jeopardy. Glaciers have been rapidly melting away from Greenland to Switzerland and from Alaska to the Antarctic Peninsula, putting more water in the ocean. More significantly, as water heats, it expands. The IPCC estimates that these two factors will cause global sea levels to rise by 1 to 3.3 feet by the year 2100. This will have disastrous repercussions for coastal communities worldwide. Some low-lying regions may be completely destroyed, including entire cities like New Orleans, Louisiana, and entire nations, like the Republic of the Marshall Islands. On top of that there's a precariously balanced store of ice in the Antarctic capable of raising global sea levels by another eighteen feet, enough to put most of the world's major cities under water.

Finally, and perhaps most serious, is global warming's potential to alter the current pattern of ocean circulation. Ocean currents play the central role in the important task of redistributing the sun's energy around the globe, acting as a sort of global conveyor belt for heat. Warm equatorial surface water is drawn towards the poles and replaced with cold deepwater flowing from the poles on the ocean bottom. Without these currents the tropics would be far hotter and higher latitudes would be much colder than they are now. One such current, the Gulf Stream, brings warm surface waters from the Caribbean to Ireland, the United Kingdom, and Scandinavia, which would otherwise have the subarctic climate of northern Quebec and Siberia. If the ocean's currents were to suddenly stop flowing, there would be deserts in the tropics and thick ice sheets over Canada, Siberia, and Northern Europe. Crops would fail, harbors would freeze, rainforests would burn. Not a very pleasant scenario.

While we needn't worry that ocean currents are going to stop altogether, global warming could cause a significant, possibly quite abrupt worldwide reorganization. There is growing evidence that very sudden reorganizations of ocean circulation have occurred in the past and may have been the "switch" between glacial and interglacial climates. As the Earth's surface warms, the key factors that drive ocean circulation are being altered.

Currents are driven in large part by differences in the density of different masses of sea water. Cold water is denser than warm water, and salty water is denser than fresh. Differences in temperature and salinity therefore cause large movements of water masses. These in turn help drive wind patterns and, together, drive global weather systems.

Global warming alters both temperature and salinity. Sea water temperatures are altered directly as the Earth's surface warms. Sea water salinity is altered at the poles, where a warmer climate may result in reduced annual sea ice. When sea ice forms it leaves a layer of very cold, very salty, very dense water underneath. This super-dense water sinks rapidly to the depths of the ocean and works its way toward the equator, setting in motion a warm water counter current to fill the "vacuum" left behind by the great masses of sinking water. These polar events—particularly the freezing and melting of Antarctica's 6-million-square-mile halo of sea ice—are thought to be the main engine driving the global system of convention currents. Less sea ice, then, could slow the entire system, upsetting the current balance.

We can only guess at the results, but they would be likely to include dramatic changes in regional climates with consequent shocks to all species, including humans. The entire ocean habitat regime could be thrown into disarray. Warm-water areas would become cold and vice versa. Upwelling areas would shut down to be replaced by others perhaps hundreds or thousands of miles away. Eggs laid in a given current would no longer float to their intended nursery. Marine life that could not adapt quickly to the reordering would perish. Stationary organisms like coral reefs would be particularly vulnerable, as would many with slow reproductive cycles.

Some scientists believe that a slowing of ocean circulation is the switch that triggers ice ages. Less circulation means greater temperature extremes, with a warmer equator and colder poles. Under this scenario, glaciers would begin forming, spreading downward from the poles and ultimately cooling things down again. If so, this is a sensible fail-safe mechanism for the Earth's climate. If conditions get so warm that the polar ice caps melt then—wham!—glaciers spread to cool things down. But what's good for the long-term health of the Earth can be disastrous for its current inhabitants. The global warming debate isn't about protecting the planet. It's about protecting humanity.

∞

THAT'S THE BIG PICTURE OF THE CRISES FACING THE OCEANS. How do these global stresses affect individual people, places, and ecosystems? In the chapters that follow we travel around the shores of the world ocean to see how the deterioration of life in the oceans is changing people's lives today.

In preparing this book I traveled nearly 100,000 miles, to six of the seven continents, to three of five oceans, to tropical and polar seas, to the shores of nations prosperous and poor. Yet despite the enormous physical, cultural, and ecological differences, I encountered the same stories over and over again. This fish was once bountiful, but we've caught them all. Marine life hasn't been the same since we destroyed those mangroves, these marshes, that once recurrent field of winter sea ice. We've put too much into the sea and are taking too much out. We built our community and culture around fish and fishing, and we fear for the future. How could we have so brazenly destroyed this incredible resource that should have provided food and wealth in perpetuity?

Marine scientists may know that the oceans are in serious trouble, but that knowledge does little good as long as most people's awareness stops at the shoreline. There are reasonable, feasible actions we can take both as individuals and as societies to protect the oceans, and these are discussed in the final chapter. But first we must understand the nature of the problem and what will be lost if we continue to treat the oceans as both a convenient sewer and an inexhaustible food resource.

What follows is the story of people and the sea—the same story, re-enacted in many different settings, of people and their relationship to the living creatures around them.

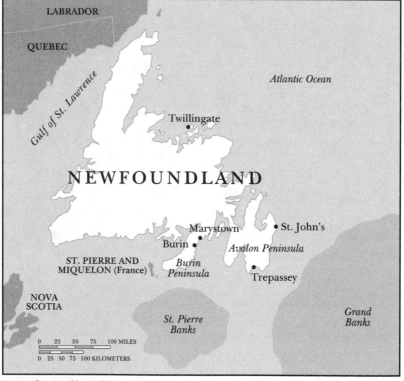

Newfoundland

three
ℜun on the Banks

 FIVE HUNDRED YEARS AGO John Cabot returned from the waters around what is now Newfoundland and reported that codfish ran so thick you could catch them by hanging wicker baskets over a ship's side.

Like most European explorers before him, Cabot had been hired by a monarch to find a westward trade route to Asia. Fifteenth-century Europeans relied on long-distance trade with Asia even more than Europeans do today, and England's Henry VII hoped the Venetian navigator would succeed where another Italian, Christopher Columbus, had failed five years earlier. Asia was the source of the spices used to preserve and flavor meats and other foods; of sugar, drugs, silk, porcelain, and cotton cloth; and of fine steel, rugs, and jewelry of a quality superior to any produced in Europe. A faster trade route would bring windfall profits to crown and country, at least until the other powers caught on. But like Columbus, Cabot was defeated by the existence of North America and the vast Pacific Ocean.

Instead Cabot discovered a resource that would change England forever, the basis of a maritime trade that would give that tiny island kingdom the wealth, skills, and shipbuilding capacity that transformed it into a global empire. He had discovered the most fantastic fishing grounds the world had ever seen, waters so teeming with life that a vast swath of the New World was colonized just to harvest its seemingly limitless bounty.

A century after Cabot, English fishing skippers still reported cod shoals "so thick by the shore that we hardly have been able to row a boat through them." There were six- and seven foot-long codfish weighing as much as two hundred pounds. There were great banks of oysters as large as shoes. At low tide, children were sent to the shore to collect ten-, fifteen-, even twenty-pound lobsters with hand rakes for use as bait or pig feed. Eight- to twelve-foot sturgeon choked New England rivers, and salmon packed streams from the Hudson River to Hudson's Bay. Herring, squid, and capelin spawning runs were so gigantic they astonished observers for more than four centuries.

Today Newfoundland's fish are gone and the seas, streams, and rivers lie quiet and empty.

☙

I ARRIVED ON THE ISLAND OF NEWFOUNDLAND just after dawn, bleary-eyed from the overnight ferry crossing from Nova Scotia. The fog sat so thickly over the port that until the very last moment it appeared that our enormous ice-breaking ship was preparing to dock on some misty afterworld. Then the sounds of automobiles and people came through the thick gray curtain of fog. Only in the last few seconds did the lip of the pier reveal itself alongside the huge, slowly advancing ship. Men in orange jumpsuits scurried out of the fog, securing lines and positioning the vessel for the birth of its substantial vehicular cargo.

It had been a routine crossing. The saloon was clearly the center of activity aboard the *Joseph and Clara Smallwood*. Entire families gathered round tables top-heavy with pint glasses and listened to the band play late into the night. Couples danced vigorous twirling Gaelic numbers while children slept in little heaps on nearby couches. The same theme recurred throughout the long night of song: A person crosses the sea in search of work, but leaves his or her heart behind on the shores of a rocky island home. The songs referred not to Ireland—though you'd think so from the music— but to another barren, beautiful, similarly sized island set against the angry North Atlantic, an island hidden somewhere in the darkness before the *Smallwood*'s bow.

The saloon was packed with Newfoundlanders. There was an elderly man who'd left "The Rock" thirty years before to find work in Nova Scotia, mining coal from tunnels extending for miles under the floor of the sea, now heading home to see his brother's family. There were three Canadian soldiers headed home on leave from peacekeeping in Bosnia. A brawny guy from St. John's, Newfoundland's capital, now worked odd jobs in Edmonton, Alberta, where he reported "they's soo many Newfies livin' that we practl'y trip over each other." His sister was graduating from high school that coming weekend and leaving Newfoundland to attend university in Ontario. It was a ship full of Newfoundlanders, but few of them seemed to still make their home there.

Bleary-eyed after the fourteen-hour crossing, they made their way to the vehicle deck, started the engines of their cars, and slowly drove over the ramp, up the pier, and onto The Rock.

Of Canada's ten provinces, Newfoundland and Labrador is the least accessible. Most of its half million people live on the great, barren island of Newfoundland. The island is 39,500 square miles, about the size of Virginia, and consists of rocky shores, barren heaths, and rolling hills of stunted pine. In winter the island is buffeted by arctic winds, and in early summer the north coast is battered by icebergs floating down from Greenland. (Labrador, the province's mainland component, is three times the size of the island but has only a few thousand residents; this stretch of exposed rock and tundra is simply too far north, too cold, barren, and remote to support a large population.) Even in summer, a trip from Boston or Quebec to St. John's entails sixteen hours of driving and fourteen hours on the ferry.

Once on the dock, most traffic headed north towards St. John's and the Trans-Canada Highway, which traverses the island. I took the deserted road south, beginning a circuit of the windswept heaths of the Avalon Peninsula. The island revealed itself reluctantly. Hills rolled beneath the car like ocean swells. Every few minutes I fell into a fog-bound trough sheltering a tiny settlement at the head of a cliff-bound cove where the only sound was the crashing of waves against the cobbled beach. There'd be a brief, sharp dip in the road and I'd begin riding to the crest above the fog. There the gray morning light revealed a boreal, primeval place where the wooden telephone poles alongside the road loomed over stunted forests of chest-high spruce and fir.

I spent part of my childhood in an equally remote and sparsely populated part of western Maine and am accustomed to the quiet pace of near-wilderness communities. Yet there was something too quiet about these Avalon shores. Too little traffic on the roads, too few people at the general store, the breakfast place by the fishing pier, the post office at lunch hour. Too many roadside filling stations were boarded up. The little hamlets clung to the fringes of harbors sheltering but a handful of boats.

I spent the night in Trepassey, a dying town on the southeastern Avalon shore. The hotel stood atop a hill opposite the cemetery and high school, behind which the town ambled down to the rocky shores of the harbor. An abandoned fish processing plant gaped open-doored across the dark blue water. Three boats were tied to the town pier: a motor yacht, a visiting sailboat, and a tiny inshore trawler without its nets. I returned up the main street at nightfall; the windows of many homes remained shrouded in darkness.

The innkeeper's family spoke with a thick brogue and were so hospitable I might have been in Ireland. "Near half the population's left in these past three years since the fish plant closed," the waitress told me. The younger people had left in search of work, and most had only found it outside the province. "The older ones are staying. If you're over forty and have a mortgage on your house or boat what can ya' do?" For his part, the innkeeper simply shrugged and said: "We're becomin' a ghost town."

Towns and villages are withering away across the province, but it's not so obvious when you're just passing through. This being Newfoundland and Canada, empty homes are cared for by friends and relatives—windows intact and unshuttered, lawns mowed if the rocky landscape allows one to grow grass. Robbery and violent crime in the U.S. sense are unknown here, and Canada's generous health and welfare programs keep Appalachian-style poverty from taking root. Here the ghost towns are well maintained.

❧

THE FOLLOWING WEEK I DROVE AN HOUR WEST from St. John's and then down the 120-mile-long Burin Peninsula on the island's rugged southern coast, which is exposed to the full impact of powerful Atlantic storms. There are a half dozen communities scattered along the end of the thick peninsula and remote from just about everyplace else. The two-hour drive down the peninsula from the Trans-Canada Highway takes you through an hour of uninterrupted wilderness—a succession of bogs and exposed granite, of glacial lakes and dense thickets of shrubbery. I drove for a full fifteen minutes at sixty miles an hour without passing a house or even an oncoming vehicle.

Burin would count as remote wilderness were it in Europe or the eastern United States, but it's fairly typical of Newfoundland. Half the province's five hundred thousand people live on the Avalon Peninsula, the easternmost appendage where St. John's and most of the early English and Irish settlements are situated. The other 80 percent of Newfoundland proper contains sparse towns whose single access roads connect with the long arc of the Trans-Canada Highway like railroad spur lines. Few roads connect one spur to another. A trip from Burgeo to Pass Island—80 miles as the crow flies—requires a 500-mile elliptical drive. But this is a vast improvement. Until a generation ago, many Newfoundland communities weren't connected by roads at all. These were the true "outports," communities settled from the sea that could only be reached by boat. Few remain now because of extensive social and highway engineering since 1949, when Newfoundland joined Canada. Until the turn of the century, when a single narrow-gauge rail line opened, it was impossible to cross the province to St. John's by land. The Trans-Canada wasn't completed until 1965. Even the larger communities have only recently acquired proper overland communications with the rest of the province. The road to Burin was first paved in 1972. Burgeo wasn't connected to the Trans-Canada until 1979.

Like so much of Newfoundland, the Burin Peninsula was founded on fishing. There's evidence that Basque fishermen used the peninsula as a summer fishing base during the early 1500s. French fishermen may have been living here as early as the 1640s, although most up and left in the early eighteenth century when the area was ceded to England. During the heyday of the Grand Banks schooner fishery, Burin towns were thriving and enormous Victorian mansions were erected in the town of Grand Bank. In the mechanized, industrial-scale, deep-sea trawler era, the peninsula was at the heart of the fish processing business, with year-round seafood plants in Fortune, Marystown, St. Lawrence, Grand Bank, and Burin proper, and seasonal ones in three smaller hamlets. Most of the twenty-nine thousand people on the peninsula either fished or worked in the plants, for companies that supplied the plants, or for the shipyard that built the offshore trawlers that fed the hungry assembly lines with massive quantities of ocean fish. The Burin plant provided a great many of the fish patties used in McDonald's "fillet-o-fish" sandwiches. It still does, but the fish it uses is imported from Europe.

That's because the impossible has happened. The last great schools of northern cod were scooped up in colossal trawler nets and the government has closed the world's greatest fishery for lack of fish: a ridiculous example of closing the barn door after the horse has escaped. In 1996 the Burin Peninsula recorded the highest unemployment rate in Canada for several months in a row. An estimated 30 percent of the workforce was jobless. "Fishin's all there was," an area fisherman told me. "Everybody got too greedy for them fish, 'en then there wasn't anything a'tall."

The night I arrived, my hosts took me for a sunset tour of the town of Burin, which is actually a conglomeration of Burin and twelve other tiny hamlets strung along Highway 220 south of Marystown. Driving across a small bridge separating two of these hamlets, Richard suddenly slammed on the brakes, waving out his open window to the driver of an oncoming pick-up truck. "Here's somebody you should talk to," he announced and hopped out of his jeep to introduce me to the driver of the other vehicle, a middle-aged, friendly faced man in glasses, a plain white T-shirt, and worn jeans. This was Ed Mayo, native son and one of the handful of doctors at the Peninsula's hospital, which could be seen on a darkening hillside nearby. Dr. Mayo was happy to speak with me right then and there, so we pulled the two trucks face to face, engines running, and stood leaning back on the hoods under the pink sky.

"Where did the fish go? All that about water temperatures, or seals, or what have you—not a bit of truth to it," he began, arms folded over his chest in a relaxed way. "Truth is the fish went through our scuppers. Tons of them. The fishermen, they're drinking friends of mine and I remember they'd tell you how they'd be high-grading or dumping the young fish overboard and it would leave a trail of dead fish as far as the eye could see. That's what happened to the fish. Greed killed the fishery."

Mayo described a region that was drying up, its people scared about the future. As in Trepassey, people in their forties and older were staying but younger people were leaving in droves. He pointed out a building up on the hill that was to have housed the region's first public swimming pool; construction was abandoned after the moratorium. At the hospital they had closed an entire wing for lack of patients, reducing capacity from eighty-five to thirty-five beds. From Dr. Mayo's perspective the worst part is an ominous shift in demand: "I do far more vasectomies now than deliveries," he says. "*Far* more." Whereas he used to receive requests to perform male sterilization from men "over forty with four kids," since the fishing moratorium he'd started getting requests from couples with only two kids. Then couples with only one and couples who had never had any children at all. "Now I'm getting single guys in their thirties wanting a vasectomy. People don't want to have children. There's this feeling of hopelessness."

The situation is even more unusual if one considers the traditionally large size of Newfoundland families. Ed Mayo is the eldest of a family of nine, and that's typical for his generation. It wasn't easy raising large families then. Mayo and his eight siblings grew up in their parents' 24-foot by 24-foot house without flush toilets or running water. He installed the house's first plumbing himself in 1967, the year he graduated from high school.

At that point Burin was so isolated that he'd been to St. John's only twice in his life. But the town was changing quickly. The government was busy moving hundreds of people to Burin and Marystown from isolated hamlets on rugged coves and offshore islands, enticing them with new infrastructure. A whole generation of Newfoundlanders made the jump from kerosene lamps, footpaths and outhouses to telephones, electricity, and automobiles, usually in the space of a single week.

Ed Mayo paused for a moment, looking out into the darkened landscape.

"You know those Indians up in the north, found themselves living on government checks. They burn down their porch. The government comes and rebuilds it. They burn it down again. Just totally lost in the new world that's been thrust upon them. Well, I think it's a bit like that here. Too many things have happened too quickly and people just can't absorb it all. Maybe we just lost our reason."

The next morning I was sitting in a Marystown office with Mike Siscoe, Director of Education for the Burin Peninsula. His school system was rapidly contracting from the bottom up. Ten years before, there were nearly 8,500 K–12 students enrolled in Burin's schools. In the upcoming 1998–1999 school year there would be 4,900. Kindergarten enrollment has fallen far more rapidly than that of the high schools, down to less than half its previous levels. In Fortune, which once enrolled 110 kindergartners each year, there were only three children registered to start classes in the fall.

"This whole peninsula was based on the fishery. Once that disappeared, the young people—the ones who'd be having small children—started leaving," Siscoe said. The community's sense of despair was compounded by memories of Newfoundland's controversial resettlement program of the 1960s, which "closed" dozens of area fishing towns. "There's a lot of resentment amongst those who remember resettlement. All these people had lived off the sea and the land with big open spaces all around them and then, suddenly they're all boxed in. Their lives have been changed so completely and suddenly."

"Now they're seeing schools closed up again. When you come along and say we're going to close the school it means to people that you're going to close the community, and they're probably right since people tend to settle where the schools are." Many people had already disappeared to the cities and prairies of Central Canada. "How many pregnant women do you see around Newfoundland?" he asked. "Not many. It's created a sense of panic, a sense that this has happened to us before."

~

THE SETTLEMENT OF NEWFOUNDLAND, indeed of much of North America, was a byproduct of the pursuit of cod. Properly dried and salted codfish would keep for long periods, an important consideration before refrigeration. It was relatively light and easy to transport. From the advent of the New World fisheries in the early fourteenth century, there was an insatiable, ready market for saltcod in Europe. It was a far cheaper protein source than beef, pork, or lamb. Catholics were prohibited from eating meat on Lent, Fridays, and other religious holidays, which made fish the only acceptable source of animal protein for 166 days of every year. Profitable, transportable, and easily marketable, cod would rival South American gold and Caribbean sugar in the New World resource-extraction free-for-all.

Seizing Inca and Aztec gold entailed a bestial, genocidal Spanish military conquest. Sugar plantations required slave labor, and millions were brought from Africa and worked to death on Caribbean plantations. The traditional cod trade only required salt, and lots of it.

France, Spain, and Portugal had ample supplies of salt, allowing them to salt and barrel fish at sea and thus largely ignore the high, barren shores of the New Founde Land. These countries focused on the offshore or Banks fishery, which could be exploited year-round, using outposts in places like Burin and Fortune for storage and the provisioning of wood, water, and other supplies but with no eye to settlement. They'd split and dry part of their catch on rocky beaches to sell on the domestic market, but the barreled "wet" cod packed offshore fetched higher prices. Most looked upon the island as a great ship anchored on the banks for the convenience of this international fishery.

England, however, lacked salt supplies and had to develop an alternative strategy to bring codfish home to market. Thus evolved the "English cure," a lightly salted, air-dried product that would eventually be favored in Mediterranean and Caribbean markets for centuries. This required extensive drying beaches ashore, called "fishing rooms," where wooden flakes were built to cure the split cod. As the trade expanded, there was extensive competition for prime "rooms," which were parceled out on a first-come, first-served basis at the beginning of each drying season. Captains soon saw the advantage of leaving a caretaker or two behind to hold down the room, protect stores and gear from winter storm damage, and prepare flakes, buildings, and gear. Unlike New England, however, the land was too poor to farm, and it was generally considered better to commute across the Atlantic, spending the winter in established and comfortable West England homes with family and friends.

For a long time the fish merchants of Devon and Dorset used their political clout to discourage any settlement of the island at all. Their concern was that settlements and local government might interfere with their unfettered access to the coastline and timber. In 1634 Parliament prohibited squatters from cutting "any wood or plant within six miles of sea shore" or occupying favorable fishing rooms. Each season, whatever English captain first arrived and set up shop in a given cove had dictatorial powers over any year-round squatters or settlers he found there.

But by the end of the eighteenth century European wars, the revolt of the American colonies, and other forces compelled the merchants to share their risks. West Country settlers erected permanent inshore fishing stations, selling fish to the merchants, from whom they purchased supplies. The settlers themselves often returned to England after a few years, but their Irish servants usually stayed, slowly populating the island. Using small open boats and hand-held lines, outport fishermen engaged in an inshore fishery. It's a seasonal fishery in Newfoundland, dependent on the cods' migration from their offshore feeding and spawning grounds in pursuit of capelin, a small schooling fish that is their favorite food. The hamlets were tiny, their size limited by the availability of adequate "rooms" to accommodate all fishers in the community. Many would pack up and go to winter homes inland to cut wood or hunt game, returning to the shore in the spring. There were virtually no roads, industry, or public services. With the exception of St. John's, the emerging colonial capital, there was no local government of any kind.

The remarkable thing is that outport life remained essentially unchanged until after the Second World War: fishermen in tiny, isolated communities pursuing cod in small, open boats to sell fish to far-off markets. Within living memory, Burin outporters were plucked from such a world and plunked into Canada's newly arrived industrial economy.

🐚

UNTIL 1949 NEWFOUNDLAND WAS A BRITISH COLONY and to this day it feels like a far-flung outpost of Northern Europe. For much of its history Newfoundland was linked to England and Ireland through family ties, commerce, political life, and trade. With the development of freezer technology on the eve of World War II America became a major market for Newfoundland cod, but there was little contact—commercial or otherwise—with Canada. It's no accident that the capital, St. John's, is located on the island's easternmost extreme, nestled between mountains on a harbor opening toward Britain, somewhere out beyond the perpetual wall of fog. From St. John's, London is closer than Calgary, and Ireland nearer than Winnipeg.

Visitors from the North American mainland might think they boarded the wrong ship and wound up on the wrong side of the Atlantic. People in many towns speak in a strong brogue, others in more of a working class Mancunian or the polished Trans-Atlantic of the BBC. The cooking is decidedly English, with fish and chips dominating menus. Until 1997 schools were still run by churches, with separate Catholic and Protestant schools in the British tradition. Some towns have annual Orangemen parades. Small town museums feature exhibits of pre-Confederation households, with framed portraits of King Edward and Princess Elizabeth in the living room and tins of English biscuits in the pantry. Other exhibits display military uniforms donated by local veterans: those of the R.A.F. and Royal Navy, or of Newfoundland regiments of the British Army whose members fought and died in Verdun trenches, on Normandy beaches, or in garrisons meant to keep Indians British.

The story of the cod's destruction begins where Newfoundland's colonial era ends. Newfoundland was settled by and for the pursuit of the cod fishery, but this was not true of Canada. To decision makers in landlocked Ottawa, Newfoundland was a backward nation in need of development and industry. The fishery was just another resource to exploit to that end. With confederation, the management of the fishing industry shifted from foggy, fishy St. John's to office warrens in far-off Ontario, setting in motion a chain of events that would ask the humble cod to underwrite the failings of a bungled, state-sponsored development plan. The costs proved too great for the cod to bear.

When postwar Britain began to set its remaining colonies free, Newfoundland had a choice to make: remain a colony, acquire independence within the British Commonwealth, or join in a confederation with Canada, that unknown land across the Cabot Straits. There was considerable resistance to confederation, particularly among the merchant class that had been running the show for centuries. The merchants and the Roman Catholic Church preferred independence, but that route had been tried before and turned out badly. Britain had granted Newfoundland dominion status in 1931, but it lasted only three years. The Depression crushed this obscure, one-industry nation. Facing bankruptcy, conditions of near-starvation, and 50 percent unemployment, the island was driven back under the British wing. Yet the St. John's establishment longed for another shot at running its own nation state.

Their nemesis was Joseph Smallwood, a journalist, pig farmer, and ardent socialist, who criss-crossed the province giving stirring speeches for confederation. Smallwood campaigned hard on the clear material and social benefits the rural majority would gain if they became Canadians. These benefits included family allowances, unemployment insurance, better pensions for the elderly, and increased benefits for veterans. It was calculated that the Canadian family allowance for a typical outport fishing household with four children would exceed their annual fishing income. Canada would take over Newfoundland's debt, its postal services, its floundering railroad, its defense, and fisheries management. Federal coffers in Ottawa would underwrite highway construction, maintain mainland ferry links, and provide substantial subsidies and block grants to the new province for many years. For a poor, struggling colony it was an extremely attractive deal.

Even so, when the votes were counted after the June 3, 1948, referendum, confederation came in second with 64,000 votes. Independence drew 69,400, and 22,300 voted to remain with Britain. Because this was not a clear majority, a vote to resolve the issue was held with only two ballot options: confederation or independence. Confederation won by less than 7,000 votes.

Confederation forever changed life in Newfoundland's far-flung fishing communities. During his twenty-three years as premier (1949–1972), Joey Smallwood attempted to reengineer Newfoundland society with Canadian public spending. Rural Newfoundland gained material comforts, social services, and an extensive social safety net. In the process it exchanged an independent, self-reliant, poverty-cursed lifestyle for a systemized, welfare-state bureaucracy controlled by distant decision makers. It traded a failed libertarian utopia for a false socialist one.

Smallwood believed rapid industrialization was the solution to Newfoundland's chronic poverty, and he focused his virtually unchecked political power on this goal. He would industrialize fishing and bring mining, energy, light industry, and pulp mills to the province, but before that he needed to do three things. First, the province would have to be knit together with roads, telephone lines, and an electricity grid (in 1949 only half of all households had electricity). Then, basic social services like schools, hospitals, and post offices would be extended to the entire population. Finally, there would need to be a concentrated labor supply in place to man the assembly lines of progress.

Smallwood concluded that the rural population had to be centralized. Three-quarters of the population lived in communities of less than a thousand, spread along Newfoundland's 6,000 miles of shoreline or on tiny offshore islands. This settlement pattern made sense during the days of the dory fishery and fishing rooms, but the advent of the marine engine and freezer technology made such dispersion unnecessary. It made the provision of electricity, transportation, and social services extremely expensive. And it was anathema to securing sufficiently large pools of industrial labor.

So starting in 1954, Smallwood organized three government-sponsored resettlement programs. By 1975 these programs had emptied three hundred outport communities, their thirty thousand residents concentrated in mainland "growth centers" like Marystown and Burin. Communities were encouraged to move by the promise of modern services and better-paying jobs combined with provincial and later federal assistance to cover the costs of moving people, belongings, even houses to the growth centers. The latter only kicked in if 100 percent of a community's residents agreed to move, although the figure was lowered to 80 percent in the mid-1960s. Thousands of homes were loaded on barges or dragged on sledges across winter sea ice to their new foundations. Practically overnight families went from outhouses, scrub boards, and wood piles to electricity, running water, washing machines, and automobiles; from eking out an independent existence on the sea to fish plant assembly lines and, later, when the growth centers failed to grow, the welfare check.

Most of Smallwood's industrial projects failed. Carpetbaggers swindled the province out of millions. For a time Smallwood relied on Alfred Valdmanis, a former Latvian finance minister and Nazi collaborator, who introduced him to German industrialists before vanishing with $200,000.* In the end, Newfoundland's unemployment remained high and its economic productivity low. By 1991, Ottawa was making transfers of more than $2 billion annually to

෴

* All monetary figures in this chapter are Canadian dollars.

keep the province afloat; without these funds life in Newfoundland would probably have been worse than before Confederation.

In fact the only industry that succeeded was the fishery.

In the early 1950s, Smallwood brought in capital to build fish processing and freezing plants in many growth centers. St. John's merchant firms switched from salt fish to frozen and fresh fish trade with the United States, boosted by that country's newfound penchant for frozen fish sticks. Small inshore fishermen began landing their catch fresh at fish plant docks in places like Marystown, Burin, and Fortune rather than shore-drying the catch themselves. Thus ended fishing as a way of life—because the entire family had been engaged in the fish "making" process. Fishing was becoming an occupation.

As Smallwood's other projects failed, Ottawa decision makers looked to the fishery as the "employer of last resort" for the struggling province. In 1957 they agreed to extend unemployment insurance (UI) to seasonal inshore fishermen who worked at least ten weeks in the fishery. To keep people employed and the fish plants supplied, the federal Department of Fisheries and Oceans (DFO) gave fishing licenses out to all comers. The number of inshore fishermen increased 33 percent from 1957 to 1964. Later DFO would subsidize boat construction, gear replacement, fuel, and equipment purchases; the number of inshore fishermen would nearly triple.

Canada wasn't managing the fish stocks, it was managing a growing welfare population. Nobody seemed concerned that fish catches were falling or paid attention to the inshoreman's warnings that the cod were being wiped out at sea before they could come into shore.

Out on the Banks, a fleet of modern industrial fishing vessels was strip-mining the northwest Atlantic. The most fertile fishery the Earth had ever seen was being turned to barren wasteland.

∞

FOR THE FIRST FOUR CENTURIES AFTER CABOT, Newfoundlanders had little trouble actually finding and catching cod. First of all there were seemingly endless numbers of them. These large, hardy, generic-looking fish are built to last. Adaptable, omnivorous, and incredibly fecund—a large female will produce 9 million eggs in a single spawning—they're remarkably successful. Atlantic cod survived in their current form for 10 million years, through ice ages and warming spells that changed world sea-levels by some 300 feet. They live twenty years or more, ensuring a diverse, adaptable breeding stock. Particularly cold spawning seasons would select cold-resistant eggs, warm seasons would bring the opposite, and with so many generations present at any one time, the cod has been able to adapt to almost everything. Everything, that is, except industrialized fishing.

Cod are part of a family of fish species known as "groundfish," so called since they generally live on or near the ocean bottom along the continental shelf. The northwest Atlantic's other groundfish include haddock, halibut, pollock, flounder, and plaice. All these species have been intensively fished and many have shared the sad fate of their cousin, the cod. But the Atlantic cod was by far the most numerous, valuable, and important of the groundfish.

Fishermen benefited from the cod's tendency to congregate in great numbers, especially during spawning periods. Atlantic cod not only school, they live in distinct stocks or populations that school and breed together. Each population moves as a vast herd from spawning to feeding grounds and rarely associates with other camps. The "northern" cod dwell off the icy coasts of Labrador and northeastern Newfoundland. Another population spawns on the nutrient-rich Grand Banks, a vast series of underwater hills sunk in shallow water off the Newfoundland coast. There are distinct stocks in the Gulf of St. Lawrence, which separates the island from Quebec and Labrador, and on the smaller St. Pierre Banks near Burin; another masses along Nova Scotia's Atlantic coast, and still more on Georges Bank off New England. These latter stocks live in somewhat warmer water and are markedly larger and faster growing than their compatriots in ice-choked Labrador. Other Atlantic cod stocks populate the European and Icelandic coasts.

When spawning, cod gather in dense clumps of hundreds of millions of fish. Northern and Grand Banks fish spawned on respective portions of the offshore banks, spawning in pairs and sowing the ocean currents with trillions of eggs. This made it possible for men to catch them in vast numbers with handlines and, in recent decades, to scoop up entire stocks with enormous nets hauled by trawlers the size of a small ocean liner. For many centuries, though, it was the cod's next move that put food on the table. After spawning, the vast schools would spread into vast sheets and head inshore, beating the ocean for prey. They would eventually find it: even vaster schools of capelin, a small open-water fish seven inches long and looking very much like a freshwater smelt. For reasons that are still unknown, some of these capelin schools spawn on the offshore banks like the cod, which gorge on them shortly after their orgies are completed. Other schools spawn inshore, actually *on* the shore. Each spring Newfoundlanders would await the arrival of the teeming millions of capelin on local beaches. They spawned in the surf in such numbers that villagers would find themselves knee-deep in a capelin soup. Whole towns gathered in impromptu festivals, scooping capelin directly into pans and frying them on the beach and gathering great piles for fertilizer or animal feed. These annual events marked the end of the long winter and kicked off the cod fishing season. For the cod were always close behind in vast schools, following and feeding on capelin, as do whales, puffins, seals, and many other creatures in the northwest Atlantic.

Cod are also greedy and will eat almost anything they can fit into their gaping mouths. In nature this makes them versatile, willing to eat whatever is available: whole mussels, crabs, lobsters, squid, even juvenile cod. It also makes baiting them very easy: They can reportedly be landed with an unadorned lump of lead, pieces of hot dog, even Styrofoam cups. Once hooked they put up no fight at all; they just hang there as they're pulled into the boat.

❧

NEW WORLD FISHING TECHNOLOGY CHANGED surprisingly little between Cabot's arrival and the end of World War II. Prior to Confederation, Newfoundland's inshore fishermen still went out in small boats that fished within sight of land, returned home at night, and stayed in port when the weather was poor. Their dories used sail-power and oars, although many were equipped with small engines. Handlines and hand-set nets were still employed, although cotton lines had replaced hemp. Fish were found by ingenuity and experience accumulated over generations and passed from father to son.

Sail power dominated the offshore fishery well into the twentieth century. New England and Nova Scotia schooners operated into the 1950s, sailing to the Grand Banks and deploying men in dories to catch the fish. Although Europe's near-shore fishery was dominated by gasoline- and diesel-powered side-trawlers, the high cost of fuel kept most of these boats from crossing the Atlantic. Up to the start of World War II, Spanish, Portuguese, and French Grand Banks fleets were dominated by sailing ships with dorymen.

Technological innovations did appear on the Banks. The French introduced a 200-ton, steam-driven trawler to the area in 1899, but it was about half the size of the schooners of the day. These new ships grew to 400–800 tons during the 1920s and 1930s, and diesel-powered side-trawlers began to appear before World War II. Despite the innovations, annual northern cod catches are believed to have remained relatively stable, at 200,000 to 300,000 tons per year between 1875 and 1945. It wasn't until after the war—during which fishing was partially suspended—that catch rates began to change dramatically.

In 1951 a strange ship flying the British flag arrived on the Grand Banks. It was enormous, 280 feet long and 2,600 gross tons, four times the size of a large side-trawler. Its superstructure, tall funnels, and numerous portholes suggested an oceangoing passenger liner, but its aft deck confirmed it to be a fishing vessel. Gantry masts supported cables, winches, and gear the scale of which nobody had seen before. Its stern was marred by a gigantic chute, a ramp from sea to deck such as whaling ships use to drag aboard the 190-ton carcasses of blue whales. But the ramp was meant not for whales but to drag onboard equally large nets filled with cod and whatever else happened to be in the water.

The *Fairtry*'s arrival marked the beginning of the end for the Atlantic cod fishery, indeed for many of the world's fisheries. She was the world's first factory-freezer trawler, a multimillion-dollar vessel equipped with all the technological breakthroughs of the war. Below decks was an on-board processing plant with automated filleting machines, a fish meal rendering factory, and an enormous bank of freezers. She could fish around the clock, seven days a week, for weeks on end, hauling up nets during fierce winter gales that could easily swallow the Statue of Liberty. With radar, sonar, fishfinders, and echograms she could pinpoint and capture whole schools of fish with chilling effectiveness.

Fairtry was in a class by herself for only two fishing seasons. Her productive advantages were clear to all fishing nations, but only countries engaged in the industrial slaughter of whales had the skills, technology, and organization to exploit the new technology. *Fairtry* was herself an outgrowth of the Antarctic whale fishery. Her builders, the Scottish whaling firm of Christian Salvesen Limited, had created her using whaling-ship technology following the realization that whaling stocks would soon become commercially extinct. Having decimated whale populations, long-distance floating factory technology was re-deployed against the world's fish. Other whaling nations leapt into the fray. Within two years of *Fairtry*'s appearance, the Soviet Union built twenty-four *Fairtry* clones. The Japanese whaling firm Taiyo constructed a fleet of factory-trawlers to ply the North Pacific pollock fisheries. West Germany, Spain, France, and Poland followed suit. The Netherlands converted at least one factory whaler into a floating fishmeal factory.

The ships grew bigger. They eventually reached 8,000 tons, towing nets with openings 3,500 feet in circumference. In an hour they can haul up as much as 200 tons of fish, twice as much as a typical sixteenth-century ship would have caught in an entire season. Recrewed and supplied by oceangoing tenders, the ships could pursue fish anywhere in the world for months on end without ever visiting a port or even sighting land. Plying international waters, they were outside the jurisdiction of the nations off which they fished. By the 1970s the Soviet Union had 400 factory trawlers on the high seas.

Japan had 125, Spain 75, West Germany 50, France and Britain 40, and dozens more were operated by Eastern Bloc nations. They plied the Georges Banks of New England, the hake stocks of South Africa, Alaskan and Bering Sea pollock, Antarctic krill, and, most of all, the northern cod off Newfoundland and Labrador.

They were strip-mining the sea.

To process fish efficiently, factory-trawlers need large numbers of similarly sized animals of the same species. Trawl nets that turned out to have improperly sized or mixed-species contents would be dumped overboard, covering the ocean surface with dead and dying fish. This practice was called "high-grading." If the trawl contained the right type of fish its contents would be sluiced down to the processing deck and sorted. Crabs, flounder, redfish, starfish, juvenile cod, sharks, and a hundred other unwanted creatures—the so-called bycatch—would be sent overboard through special discard chutes; for every three tons of fish that were processed, another ton or more of other creatures were killed in this manner.

To get large numbers of adult fish, captains pinpointed the massive spawning congregations of cod. Scores of factory-trawlers would converge on a spawning school, scooping them up before they could reproduce. Mid-water trawling was soon perfected, allowing factory-trawlers to mine herring, capelin, mackerel, and other small fish that cod, whales, and other animals rely on for food. Spanish pair trawlers—two ships towing one incredibly massive net—swept vast areas of the ocean clean, discarding large proportions of bycatch.

At night the Grand Banks often resembled a city, such was the light cast from the vast fleets of factory ships as they devoured all the sea had to offer.

In 1968, the cod catch peaked at 810,000 tons, almost three times more than had been caught in any year prior to the *Fairtry*'s arrival. Then, despite increased effort, larger nets, more accurate fish finders, and larger on-board processing plants, total cod landings fell.

Two Canadian fisheries scientists, Jeffrey Hutchings and Ransom Myers, have calculated that about 8 million tons of northern cod were caught between Cabot's arrival in 1647 and 1750, a period encompassing twenty-five to forty cod generations. The factory trawlers matched that take in only fifteen years, well within a single cod's lifetime. The trawlers were scooping up fish many times faster than the ecosystem could replenish them. Not just cod but also other groundfish, including flounder, halibut, and haddock, were decimated. The limitless fish stocks were on the verge of collapse.

Meanwhile, fewer and fewer fish survived to migrate inshore, where frustrated Newfoundland fishermen waited in their small boats. Inshore catches fell by two-thirds between 1956 and 1977. The growing Canadian fleet of 50- and 60-foot steel trawlers—so-called mid-shore vessels—couldn't compete with the massive behemoths on the Banks. Fish plants were failing. The province was in an uproar. Something had to be done.

In 1977 Canada followed Iceland in unilaterally extending its territorial waters from twelve to two hundred miles offshore. Foreign factory-trawlers were kicked off the Banks, except for a small area called "the Tail" that lies beyond two hundred miles. But by this time the groundfish stocks were so depleted that many factory-trawlers had already moved on to strip-mine elsewhere.

Still, the decision was greeted with euphoria in Atlantic Canada. Finally the Banks would be used for the benefit of Canadians. With the area now under the sovereign control of a single nation, professional fisheries managers could ensure that the fish would recover and form the basis of a new, thriving Newfoundland economy.

Or so everyone thought.

In a remarkable display of shortsightedness, Canada proceeded to build a deep-sea trawler fleet of its own. Foreign fishing had shattered the ecology of the northwest Atlantic fisheries. The Canadian government proceeded to finish off the survivors.

In the late 1970s most people still clung to the delusion that the ocean's bounty was infinite. The foreign fleets had clearly damaged fish stocks, the reasoning went, but with proper management there would soon be so many fish we won't be able to catch them all. This illusionary "fish surplus" was seen as a panacea for Newfoundland's long-depressed economy, which continued to drain tens of millions of dollars from the federal treasury each year. The foreign fleets had been greedy, yes, but in responsible hands the new technologies they had used would provide the basis for new prosperity across Atlantic Canada. All that was needed was a team of fisheries managers and a massive capital infusion, and everything was going to be great.

The new policies ignored the fact that the province's traditional inshore fleet already suffered from drastic overcapacity. Encouraged by the special Unemployment Insurance program and a range of fishing subsidies, thousands of Newfoundlanders had joined the fishery at the same time inshore stocks were collapsing. This trend continued after the 200-mile limit was declared. The number of inshore fishermen grew from 13,736 in 1975 to 33,640 in 1980.

Rather than addressing this overcapacity, federal and provincial officials supported the creation of Canada's first deep-water trawler fleet. These were large, modern vessels, typically 150 to 180 feet in length, capable of fishing the grounds where the factory trawlers had prowled. Most were owned by large corporate processing firms, which constructed new fish plants in preparation for the expected bonanza; Atlantic Canada's fish processing sector grew by two and one-half times in the late 1970s.

The expansion of the domestic industry created an economic imperative that more fish be caught. "Under-utilized" fish stocks had to be captured to keep processing plants busy. So while the new fleet was under construction, joint ventures were set up with foreign factory-trawlers to capture fish on the banks; the trawlers would land part of their catch at Newfoundland fish plants and keep the rest to land at home. Such arrangements had the approval of the Foreign Ministry since they helped mollify some of the foreign governments angered over Canada's controversial seizure of most of the Banks.

Foreign policy considerations resulted in many foreign trawlers being given quotas within the 200-mile limit, a move extremely unpopular with inshore fishermen. Perhaps the most egregious of these was Ottawa's granting to the USSR a 266,000-ton quota of offshore spawning capelin in 1978. The Soviets were happy to oblige: They had severely restricted the capelin catch within their own waters after the stocks collapsed from overfishing. Nobody questioned the wisdom of aggressively fishing the keystone species of the ecosystem.

Other factory-trawlers converged on the Tail of the Banks, just outside the 200-mile limit. Spanish trawlers often utilized small mesh nets or net liners that resulted in enormous bycatch of juvenile cod and other species. Others plied for herring, mackerel, and other fish.

The collapse of the Banks was right around the corner.

∞

ON THURSDAY MORNING the sea tucked Burin into a cool, briny blanket of fog. I felt my way down a maze of narrow, winding roads to Burin's fishing museum, which occupied both a three-story clapboard house and a newer building across the road that once housed a bank branch office. Donald Paul, a thin, bearded man in a flannel shirt, greeted me shyly and welcomed me inside. He led me up the stairs past exhibits of fishing gear, schooners and trawlers, cod and capelin, the endangered and extinct ingredients of Burin life. "I've one of those stored away at home," he said, pausing at an exhibit of a nineteenth-century cod trap. "Old one got tore up by a whale."

Donald Paul has been fishing the waters off Burin since 1974. He's from a Burin family as large as Ed Mayo's. As he puts it, "we never had any trouble going out and starting a game of ball," although his own kids can't find enough peers to make a team. He's an inshore fisherman, the self-employed sort of fisher of the popular imagination. He owns his own small boat and works the near-coastal waters around Placentia Bay, landing fish ashore at the end of the day. He's lucky to still be fishing. Of the twelve groundfishing zones around Newfoundland, only the one that includes the waters off Burin is still open, although fishing has been severely restricted. Elsewhere groundfishing is banned entirely.

"Back when I started there were plenty of fish. There were no restrictions so you could start whenever you'd like, but it was usually in mid-April, late April, beginning of May," he recalled while our tea was steeping between us on the backroom table. "Oh, it was cold that time a' the year. I seen times where I almost cried my hands were so cold." Wearing thick gloves in the hook-and-line fishery was too dangerous. If your gloves snagged on a sharp hook during the deployment the weight of the others could drag you overboard. You didn't stray far from shore in those days. Electronic navigation was out of the reach of the average inshoreman.

Don fished through the summer and collected unemployment for the winter months. For a couple of years he crewed on the new deep-water draggers that fed Burin's fish plants year-round, but says he didn't like it so well. "Too much time away from home and I wasn't used to it. You'd be out two weeks or twelve days, home for two days, then back out again, all year 'round. Only break you'd have was Christmas. You'd come in on the 23rd, 24th of December and go out again on the 28th." At sea, the fish plant trawlers ran day and night, the men working eight-hour shifts alternating with four hours in their bunks. Don worked on deck, gutting the mountains of fish dumped by each net and sending them down to the hold. "Wasn't my kind of fishing," he said.

Some tourists in the next room were gawking over old photographs of cod traps, surprised that "in the olden days" you could catch cod by stringing a fence out from the shore; in hot pursuit of capelin, the cod would steer along the fence to the end where they found themselves in a box-shaped cod trap.

"I'd say the first year I noticed something was '78," Don was saying. "Normal years we'd get 200,000 pounds of cod, but that year it was more like 70,000 pounds." Burin fishers stalked the seas, but their catch stayed at these low levels for five years. The foreign draggers were gone, so what was going wrong? In the mid-1980s Don remembers a couple of good years. "Then all of a sudden they just crashed, '89 I guess it was. Cod come in from offshore following capelin, but the capelin didn't come, so neither did the cod."

Talk around Burin was that the Soviet factory-trawlers were wiping out the capelin just as the foreign fleets had done to the cod. Only this time they were doing it with the blessing of Canada's Department of Fisheries and Oceans.

For as long as anyone could remember, the capelin had come ashore to spawn on the second to last week of June, exam week in local schools. "Oh, they'd be everywhere on the beach. Stay a couple of weeks. Just pick 'em up and pan fry them or store some up for the winter. They'd be all over the place right out here, over across the other side of the bay, over at St. Lark's, Red Harbor, all along the way."

Meanwhile the deep-water trawlers kept coming in, unloading cod by the hold-full on the fish plant dock a few hundred yards from the museum. Only the fish they were catching kept getting smaller and smaller. The offshore fleet continued to catch 300,000 tons a year of cod, but they were catching many times more individuals: adolescent fish three, four, or five years old that had yet to begin spawning. They were catching the very fish on which a stock recovery depended.

"I think the biggest story was that people was greedy. I mean everybody," Don said. "Lots of fish was coming in and there were more fishing licenses and bigger boats and people was making money at the plants. There wasn't anything else 'sides fishing, so everybody turned a blind eye to what was happenin' right in front of 'em."

This year Don hadn't seen a single capelin. Not one. In Burin the fish had simply stopped migrating ashore.

                             ☞

THE SECURITY GUARD WAS LOST. We were on the right floor but had turned into the wrong sub-warren of offices. We followed the labyrinthine corridor, twisting past identical office doors, a half-empty water cooler, and a wall chart of Canadian fishing grounds, only to be confronted by a cul de sac. We were trapped, cod-like, somewhere in Department of Fisheries and Oceans bureaucracy. Finally a secretary came by. "Groundfish division? Oh that's, well, actually, I'll take you there. It's easier than trying to explain."

When I finally arrived, Bruce Atkinson welcomed me into his office in a light Newfoundland brogue. His desk was covered in reports, surrounded by shelves stuffed with fisheries management reports, data sets, textbooks, and policy position papers. Atkinson, middle-aged and bespectacled, was the division head for groundfish at DFO-St. John's, where he has worked since before Canada declared the 200-mile limit. His nondescript office chair was, in fact, a hot seat like no other. For while the capelin crashed and the inshoreman began panicking, DFO had maintained that cod stocks were growing. For a decade following the eviction of the foreign fleets, DFO recommended quotas that the industry—try as it might—couldn't fill. DFO had woefully underestimated the damage factory-trawlers had done to the ecology of the northwest Atlantic. Until 1989, DFO thought there were more than twice as many spawning cod in the ocean than there actually were.

"Safe to say it was as much of a surprise to us as to anyone else when we realized the truth," Atkinson said. Although Canada had one of the most sophisticated management systems in the world, the state of fisheries science knowledge was woefully inadequate to understand, better yet conserve, the northern cod and their ecosystem.

To this day the discipline is dominated by statisticians rather than ecologists. Each species was managed separately, as if its well-being were unrelated to that of the creatures it ate, competed with, or was eaten by. Fishing quotas were based on polling and demographic modeling techniques, computer models, and logarithmic calculations. Fish were managed not as wildlife but as if they were corn or soybean futures. DFO would input statistics on reported commercial landings and the composition of annual survey trawls and output recommended quotas for each species based on what computer models suggested were sustainable levels.

But when Canada started managing the stocks in 1977, DFO scientists didn't have long-term landing data, scientific survey results, or reliable models. They didn't even have computers.

"Back then we were running our calculations with calculators," Atkinson recalled. "Then in the late seventies we had these HP–97s which were big boxes that took magnetic strip cards and those were the cat's meow since you could actually do Virtual Population Analysis. But you couldn't come to the assessment meeting with five or six runs since it took so much time. You did it once and went with it.

"The Canadian fleet was just getting involved in the offshore and we were just learning from them about catch rates. If those increased we had no way of knowing if it was because of improving technology, something in the environment, or that there were just more fish." There was a lot of guess work and, despite best intentions, the scientists hadn't guessed right.

The shock came in 1988. New modeling techniques and the latest stock survey revealed that many groundfish stocks were on the edge of collapse. The northern cod stock—by far the largest and most important fish resource in Atlantic Canada—was in the worst shape of all. Atkinson and his colleagues concluded that quotas had to be more than halved to prevent this stock's collapse. Politicians were appalled. If implemented the new quotas would cause economic chaos throughout eastern Canada. So the politicians compromised on what could not be compromised. Quotas were cut by only 10 percent.

More frightening data poured in, confirming that the stock was in serious trouble, that fishermen had been capturing as much as 60 percent of the adult cod every year for several years running. Ottawa's 1990 quota was trimmed from 235,000 to 199,262 tons, still well above the 125,000 quota fisheries scientists had recommended the year before.

Plants closed in Trepassey, Grand Bank, and elsewhere. Two thousand people were out of work. Canada released $584 million in emergency assistance. Fishermen tried as hard as they could, but could only catch 122,000 of the 190,000-ton cod quota for 1991.

The stock was in free fall. New DFO research vessel surveys revealed that the estimated mass of spawning northern cod—about 1.6 million tons when the *Fairtry* arrived—was down to a mere 130,000, a drop of 92 percent. Unable to face reality, Ottawa set a 1992 quota of 120,000 tons.

A few months later the spawning biomass of northern cod was revised downwards to a shocking 22,000 tons, a drop of 98.9 percent from its 1962 level. Smaller stocks on the minor banks of Newfoundland had fallen by 77–95 percent of historic levels. Politicians finally realized that regardless of what quotas they set, nature had spoken: There would be no fish to feed the plants and working families of Atlantic Canada. In July 1992 the government finally did what it should have done four years earlier: closed the Banks altogether to allow the stock to recover.

But by then it was far too late.

❧

FROM FORTUNE I BOARDED A TINY PASSENGER BOAT and crossed 20 miles of rolling, overcast sea to France.

The tiny islands of St. Pierre and Miquelon are all that remains of France's North American empire, which once included most of Canada, the Mississippi Valley, and the Upper Midwest. France lost its hold on Newfoundland, Quebec, and the Maritimes after a series of defeats in New and Old World wars. But these grassy, fog-bound islands—93 square miles all told—remain under Paris's control. Today they are not a colony but a full-fledged department of mainland France, much as Alaska and Hawaii are integral to the United States.

In fact, the town of St. Pierre is more French than France itself. A comfortable huddle of pastel-colored frame houses and narrow winding streets, it looks as if it had been transplanted from rural Brittany for protection from the taint of European integration. At the dock, the immigration booth was protected by the men of the Gendarmerie Nationale in their dark blue *kepis*. Outside, near the Place du Général de Gaulle, locals played *petanque* with polished steel balls while shopkeepers did a brisk business in croissants and pain au chocolat. Satellite dishes draw in TVF and Radio France, sparing Pierrais from exposure to the CBC's English and—*plus horrible!*—Québecois French. Everything is paid for in francs, although

I spotted an old man grimly reviewing an EU poster depicting the soon-to-be-released Euro.

St. Pierre and Miquelon's six thousand people were even more dependent on the sea than Newfoundlanders. Virtually everyone here is at least a fourth or fifth generation descendant of fishing families from Normandy, Brittany, and the French Basque country. Their great, great grandfathers came here to fish cod, and if it weren't for the collapse of the stock, that's what Pierrais would be doing today. There's nothing else: no industry, no agriculture, no timber or mines; only a trickle-trade of language students and bootlegged booze with Canada. But unlike Newfoundlanders, Pierrais don't just up and leave.

Hubert, my guide on the island, explained, sipping cappuccino in the seat of his idle van. "This is a small place, a comfortable place where everybody knows everybody and everything they have ever done. Paris isn't like that. Europe isn't like that. The young people may try it for a while, but it's uncomfortable for them and they come back to St. Pierre. If you grow up on St. Pierre then St. Pierre is your home all your life."

The reason Pierrais aren't compelled to leave—now that the European Union has imposed a moratorium on St. Pierre's portion of the Banks—is because France is willing and able to underwrite the entire economy. Thirty-five percent of adults work for the government, administering life for the rest. Another 40 percent worked in the now defunct fish processing plant and idled trawler fleet; technically they still do. Although nobody has worked at the plant for years, Hubert says, Paris continues to pay salaries and benefits as if they did. The island is truly a welfare state.

"People hunt and fish, pass the time with friends, take walks, go on vacations," Hubert said. "Sometimes it's a little boring, eh? But, you know, it's a good life."

For Paris the prestige and strategic value of St. Pierre far outweigh the cost of keeping six thousand people alive and well in the midst of a fished-out ocean.

Newfoundland doesn't have this option.

❧

MARYSTOWN HAS THE SLAPPED-TOGETHER FEELING of many Smallwood-era growth centers. There's a traditional Newfoundland fishing town along the shore, a series of cookie-cutter, prefabricated subdivisions on the hill, and an enormous shopping mall plunked down in between, flanked by the shipyard and the peninsula's only traffic signals. Government offices are housed in a scattering of small, kit-built office buildings, a public sector deployed as an afterthought four hundred years after the town was settled.

Keith Osborne's office was in one such building. A native of the Burin Peninsula, Osborne had been involved in community development for fifteen years, since the heyday of the Canadian deep-sea trawler fleet. Now in charge of small business lending at the Burin Peninsula Development Corporation, Osborne's task is growing harder by the year.

"Those who worked on the offshore trawlers in Marystown and Burin did fairly well and owned their homes outright. If you own the roof over your head you're going to be able to weather hard times easier," Osborne was saying. Homeowners were sticking it out, holding on however they could. Fishermen switched over to the new snow crab, shrimp, and scallop fisheries. One man started a mussel farm, another makes cabinets. The fish plants in Burin and Fortune's fish plants are open for part of the year, importing whole fish to provide seasonal income for several hundred. Then there's unemployment insurance and low-wage work at the mall, whatever it takes. But the young and mobile are voting with their feet.

"What you see now is that young people are moving away with their children, hoping for work and prosperity in Alberta or wherever they're going. It takes a toll on the ones who stay behind," Osborne said. "It casts a pall to see your neighbors or brothers or sisters going away. And a whole lot more are going away than are coming in."

Between the 1991 and 1996 censuses, Newfoundland's population dropped by eighteen thousand. This was a drop of 2.9 percent for the province as a whole and represents only the beginning of the expected demographic change. For Canada has been spending billions of dollars in special assistance to help those who've lost work because of the cod moratorium. That money will end very soon, given that the cod have shown no sign of recovery more than five years after the moratorium was declared. Only when this program ends will the full brunt of the socio-economic damage become apparent.

The crisis put thirty-five thousand fishermen and plant workers out of work, and over half of all Newfoundlanders who reported having taxable income in 1994 collected unemployment insurance. The provincial economy shrank by 4 percent in 1995 alone, while Canada's other nine provinces grew. Despite nearly a billion dollars in federal "equalization" payments, Newfoundland faced a massive budget deficit. Over two thousand public employees lost their jobs. Grants to Memorial University of Newfoundland were cut. New income taxes were placed on those earning $60,000 or more.

Peter Sinclair, a sociologist at Memorial University, conducted surveys of the residents of typical fishing communities along the Northeast coast. In 1994, 75 percent of those under twenty-five expected to be living somewhere else in five years' time, as opposed to 7 to 8 percent of those in their fifties and sixties. Sinclair foresees an adjustment period "of five to ten years marked by extreme stress in rural Newfoundland and a depopulation of at least some parts." Then, perhaps, resourceful Newfoundlanders will have forged a new way of life. But as Sinclair says, "I would be a little scared if I was a person with few resources living in a rural part of the province."

One thing is clear: The cod are not coming back.

❧

AS STOCKS COLLAPSED, THE COD MORATORIUM expanded to include eleven populations. The closures—which included many other groundfish—extended from northern Labrador and the Newfoundland Banks, down the Nova Scotia coast, and on over the American border to Georges Bank, where American managers had repeated the mistakes of their Canadian colleagues. In five years, politicians said, stocks will recover. Newfoundland will have a third chance to get it right.

Six years later the northern cod shows no sign of recovery. None at all. Scientific surveys have revealed the most horrid fact that not a single generation of juveniles has survived to young adulthood (about age three), ensuring that there can be no new spawners to help restore the stock, as cod don't reach sexual maturity until age six or seven. The stock remains "at or near its historic low" of 22,000 tons of adult spawners, 1 percent of its pre-*Fairtry* level.

Even if left alone, the northern cod may never recover. Industrial technology and human greed may have so decimated these hardy fish that they can no longer hold onto their ecological niche. The crash could be irreversible.

"They might never come back, at least not in their former abundance," says Richard Haedrich, a fisheries scientist at Memorial University. "Once you start changing the whole ecosystem, the community structures and sizes, you've got a whole new ball game."

Haedrich and his colleagues have been uncovering staggering changes in the ecology of the offshore banks and paint a picture of a system in distress. Cod are at an all-time low, of course, as are stocks of other commercial fish species like redfish, halibut, and plaice. But so too are fish species that have never been targeted by fishing fleets. Haedrich found that certain wolffish, eelpouts, and grenadiers—creatures too sparse, ugly, and unpalatable to have ever attracted fishermen—have experienced dramatic crashes in their populations. The average size of almost all bottom-dwelling fish has also fallen to a fraction of their 1979 levels. The mean size of cod fell from over two kilos to less than one, an untargeted grenadier from one kilo to 400 grams.

"Let's say you came back here after living in San Francisco for twenty-five years and noticed that everyone was only two and a half feet tall," Haedrich said, pointing at colored wall charts demonstrating crashes of everything from flounder to wolffish. "You'd say 'Hey, what's happening here?' Something very dramatic is taking place in this population."

It's probably no accident that populations of these and other un-wanted species have crashed in parallel with fished species. These fish share the bottom environment with cod, haddock, and flounder, and large numbers were likely scooped up in gaping trawl nets as bycatch. Their crashes went unnoticed since nobody's living was di-rectly linked to their abundance. One large ray—the barndoor skate—may now be extinct since the sixteen-square-foot bottom-dweller was easily ensnared in groundfish trawls; its estimated num-bers fell from six hundred thousand when the *Fairtry* arrived to five hundred in the 1970s, the last time that anyone actually saw one. Amazingly, nobody noticed the skate's absence until 1998. "It's as though the bald eagle disappeared and no one noticed," says Ransom Myers, the Canadian biologist who exposed the species' virtual extinction. Many other commercially uninteresting species may have shared the barndoor skate's fate, quietly vanishing from the face of the Earth. With every species that is extinguished or dec-imated, the resilience and potential of the living community of the ocean floor is reduced a little more.

There is growing evidence that the trawlers may not only have scooped up all the fish but also laid to waste the entire seafloor en-vironment those fish required to survive. In the late 1990s marine scientists began assembling evidence that modern fishing gear causes massive physical and ecological disturbance to the benthic community. The continental shelf—where most ecological and, thus, fishing activity takes place—is not a featureless plain of muds. Rocky outcroppings, boulders, cobbles, and pebbles provide "struc-ture" on and around which living communities can thrive. As on land, plants and animals themselves create structure and habitat for other creatures. Algae, worms, brachiopods, and bryzoans form shell-like structures on rocks and pebbles, stable oases amidst shift-ing mud and sand. Generations of mollusks form massive shell beds to similar effect. Sponges, anemones, sea pens, corals, and algae form vertical structures, which other creatures attach to or hide within for protection. Juvenile cod and other fish depend on such structures for their survival. Here they can hide from predators and find small crustaceans, crabs, and other creatures to feed upon. All benefit from the chemical cycling of seafloor sediments by a wide variety of worms, just as healthy gardens (and the people who eat from them) depend on the unseen efforts of earthworms.

Modern bottom trawls destroy these structures like gigantic plows. Dragging the bottom for cod or flounder, nets are spread open by a pair of metal "doors" or "boards" weighing tens to thousands of pounds. The bottom of the trawl mouth is a thick cable bearing the weight of 50- to 700-pound steel weights that keep the trawl on the seabed. Many drag tickler chains along the bottom to scare shrimp or fish off the bottom and into the net. Scallop, oyster, and crab dredges consist of steel frames and chain-mesh bags that plow through the seabed to sift out target species. With each pass, trawls and dredges overturn, scrape, or sweep away boulders and cobbles; crush or ensnare bottom plants and structure-building animals; and kill or disrupt worms and other animals in the sediment. Most species take months or years to reestablish themselves; some take decades or centuries. None are given that much time.

Modern fishing vessels are so numerous, efficient, and seaworthy that, on average, a given portion of an offshore fishing banks is revisited *every four to twelve months*. In a 1998 paper, Les Watling of the University of Maine and Elliot Norse of the Marine Conservation Biology Institute likened trawling's effect on the seabed to that of forest clear-cutting, except that it occurs over an area of the Earth's surface that is 150 times greater. The factory-trawlers may have destroyed so much juvenile cod habitat that the Banks are no longer capable of nursing large numbers of the fish. Recovery would require decades without trawling.

These disruptions have allowed opportunistic creatures to move in. In some areas small skates and dogfish (a small shark species) appear to have taken over the cod's niche in the ecosystem. Scavengers like the snow crab and American lobster underwent incredible population explosions as the cod stocks collapsed. Large cod once ate these crustaceans, but now there weren't any cod large enough. There's some evidence that the crab and lobster booms were also fu-

eled by the huge quantities of dead animals falling to the seafloor after being dumped as bycatch from trawlers. Shrimp—once rare in
the area—have also undergone a population explosion. Now there
are so many crabs, shrimp, and lobster that Haedrich thinks they
may be capturing much of the ecosystem energy once consumed by
cod. In other words, the Banks ecology is now growing snow crab
and lobster instead of—to the exclusion of—Atlantic cod.

"Do you want a crab fishery or a cod fishery?" Haedrich asks.
"You can't have both—not at these levels."

The fishing industry might opt for crab.

Here's the great irony: In terms of total value landed, Atlantic
Canada's fishing industry is thriving. In fact, it's never been better.

In the midst of the moratorium Newfoundland's fishing industry
has been breaking records. Not in terms of total fish landings, of
course; those had collapsed with the cod from a provincial high of
550,000 tons in 1988 to 135,000 in 1995. But the value of the 1995
catch—$345 million—was $55 million higher than 1988, the best
year on record. More than 90 percent of that value was from shellfish, 51 percent from snow crab alone. Crab and shrimp are now
king, lobster a distant third.

But these invertebrate crustaceans cannot replace groundfish's
economic and ecological roles. Shrimp are an offshore species re-
quiring large, expensive deep-water trawlers to harvest; currently a
handful of these trawlers capture virtually all of Atlantic Canada's
shrimp harvest. Neither species requires extensive processing. Fresh
lobster and snow crab are sold live, shrimp whole or, when frozen,
cooked and beheaded. The shellfish boom is shared by relatively
few. Sixty percent of the employment in the sector has been lost,
most of that in processing plants. Shrimp and snow crab can't provide the basis for a way of life.

The shellfish boom may also be short-lived. Marine ecologists
warn that snow crab, lobster, and shrimp are boom-and-bust
species, temporarily thriving in the absence of competition and predation by groundfish. Eventually the ecology will restabilize—albeit
in a simplified, less resilient form—and the invertebrates will cease

their rally. Like the stock market, the shellfish stocks of the 1990s may be afflicted with "irrational exuberance."

Nor is it certain that Canada has learned from its mistakes with the cod. The fishery has simply turned to alternative species farther down the food chain and, in at least some instances, may be pushing their populations towards collapse. After several years of intensive fishing, total landings of "under-utilized" fish like herring, eel, and skates dropped significantly in both Newfoundland and Atlantic Canada as a whole. Thousands of tons of lumpfish are harvested for their roe and urchins for their gonads—both products prized on the Asian market. Despite strong fishing effort, lobster catches in Atlantic Canada fell by nearly 20 percent between 1991 and 1994 and continue to do so; Newfoundland fishermen now catch about 85 percent of the fishable stock every year, leaving very few spawning females to replenish the population. An advisory body to DFO noted in 1996 that if there were no fishery, lobster egg production in the region would be 98–99 percent or more than it is today.

Meanwhile the keystone species of the entire ecosystem—the humble capelin—has again become the target of a sizable fishery. The pattern mirrors that of the cod before them. In 1990 Newfoundland fishermen captured 138,000 tons of the tiny fish, less than the quota that had been set by DFO. The following year they were unable to catch even a third as much, although DFO had only lowered the quota slightly. Late as usual, DFO closed the offshore fishery in 1992 and cut inshore quotas; again fishermen were unable to attain but a tiny fraction of their reduced quotas since female fish were too small to market. The rules on marketable size were jettisoned and fishermen caught 24,000 tons of capelin in 1996, still less than the quota set by DFO. DFO maintains that the stocks are healthy; rural residents across the province think DFO's models will once again be proven wrong.

☙

TWILLINGATE LIGHTHOUSE IS PRETTY MUCH the end of the road. It's only a few miles west of Fogo Island, one of the four corners of our planet according to the Flat Earth Society—continue from there and you'll fall off the edge of the Earth. It certainly feels like it. From the Trans-Canada Highway one drives two hours north through twisting roads, past nesting eagles and stark, iceberg-scoured shores, over causeways and bridges linking a handful of near-shore islands to the "mainland" in an irregular chain. Twillingate Island is a half a dozen islands removed from the shores of Newfoundland proper. Drive through town past the Iceberg Boat Tour companies and out along the high, barren cliffs of Long Point. The road ends beside the lighthouse, perched high above the frigid ocean atop sheer, north-facing cliffs, a lonely sentinel on windswept shores.

In the yard I met a tall, bearded, fit-looking man carrying pails in each hand. This turned out to be Jack May, one of North America's last remaining lighthouse keepers, a Twillingate native who has kept Long Point Light lit and maintained for two decades. May is also an amateur poet. He's been a regular feature on CBC Radio's provincial morning show, reading rhymes on pigs, politicians, and Newfoundland history. One begins:

> A tourist from the USA
> The other day to me did say
> "Hey man where's all that poverty
> We hear about on our TV?
> You all have no Cod fishery
> But those new homes, new cars I see,
> Tell me that all you Newfies be
> In better financial shape than we,
> From that great land where all are free
> And live in total luxury"
> I said, "You'll find no poverty
> In Newfoundland here by the sea,
> 'Cause we have finally been set free
> From the Fishery Economy,
> In Chretien and in Wells we trust
> Our leaders they look after us,"

So our people all said, "What the heck
We'll just wait for a government cheque,
As we all bask in the sun and rain
And wait for Cod to come again."

In person, May wasn't optimistic about the cod returning. "We don't seem to be able to see the big picture, you know. We see a few extra shrimp in the system and we go like Hell after them and grab them all up and say, 'Hey, we made a lot of money on that! Here's a fishery worth billions!' But it's only going to stay that way if we look at it ten years down the road and see what Mother Nature's ground rules are for what we can take. We don't set the rules, we can only try to work within them."

An elderly tourist couple were coming down the spiral stairs to the tower above. May asked them if they'd seen the light. A knowing smile came to the woman's face. "Dear, I saw the light years ago," she said. "Hallelujah," May offered, before turning back to me.

"My non-experienced and non-professional opinion is that the whole cycle of nature is completely screwed up," he continued. "Beaches used to be littered with capelin so thick you could walk on them. Everything followed them: salmon and cod and the fishery. Puffins, seals, whales—capelin's what makes them reproduce. We don't see inshore capelin anymore.

"When I grew up in the thirties and forties all we took capelin for was to put a little fertilizer on the ground so vegetables would grow and dry some more for the winter. But now they have this roe market. I think its strange since if you've ever seen caviar and feel the way I do about it, it ain't something you're goin' to rush out and buy. Tastes like salty fish eggs. But seems if you eat caviar then you're someone in the world. Now they're shipping females to Japan to make their roe into caviar. Some cultures have this warped impression that if you eat the basis of life—which is the egg—you'll live longer. But basically they're taking all the females out of the system, so one day it's possible there won't be any."

I bid Jack May farewell and walked out onto the bluffs in front of the lighthouse. August, warm and pleasant, turned cool and damp in the offshore breeze. The edge of the Earth spread out before Newfoundland, broken by a solitary tabular iceberg drifting down from the Arctic. I watched for a while, but couldn't tell if it would come ashore or melt away unseen into a vast, empty ocean.

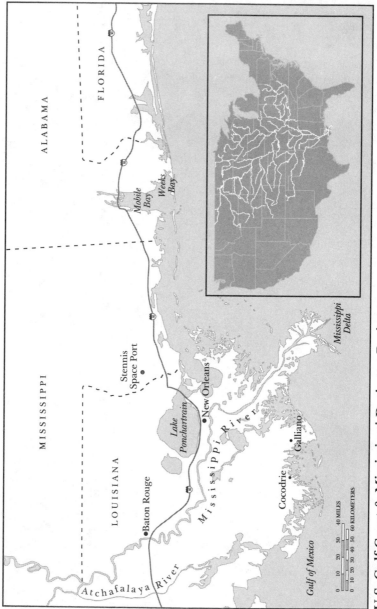

U.S. Gulf Coast & Mississippi Drainage Basin

four
Muddied Waters

RIVER ROAD IS A NARROW RIBBON of asphalt winding alongside the Mississippi River for 90 miles between Baton Rouge and New Orleans. Actually, it's two roads, one on each side of the mile-wide river, connected by the occasional bridge or ferry. These run through one of the strangest human landscapes in America, a chain of rural parishes that bring together the nation's great shortcomings: the legacy of slavery, the gap between rich and poor, and the cruel greed of some of its largest corporations.

I traveled River Road for the first time on a quiet Sunday in January, a time when southern Louisiana still basks under a balmy subtropical sun. I got on the interstate, a most impressive network of infrastructure. It whisked cars and trucks up and over the high levees protecting New Orleans from the twin flood threats of the river and the hurricane-prone Gulf of Mexico. It carried them on stilts for miles above uninterrupted cypress swamps and over the muddy, brown Mississippi before gently setting them down on the spongy eastern bank. I took the first exit and—180 degrees later—arrived in another time, another world.

The first few miles were not unusual. I passed through fields and small stands of wood, modest wooden homes with aging pick-up trucks in the driveway, sleepy main streets of one-street towns. Most of these were on the left-hand side of the road. The road's other shoulder tended to hug the base of a steep, three-story tall, grassy ridge that stretched, virtually uninterrupted, for more than a thousand miles from the mouth of the Ohio River to the mouth of the Mississippi a hundred miles behind me. These are the levees, mighty earthen defenses with which humans are trying to domesticate one of the world's largest and most powerful rivers. There's a twin on the opposite side of the river, alongside the "other" River Road. In Louisiana the twin levees are said to be the largest man-made structure on Earth, longer and thicker than the Great Wall of China.

In fact the levee is so large that it completely conceals the river. To catch a glimpse of the river you must drive up one of the little dirt roads that lead to the levee's crest every half mile or so. With some trepidation I negotiated my rented compact up the enormous barrier to the crest, throwing up a cloud of dust in my rear view mirror. From on top of the levee I confronted a confusing vista. On the other side of the levee a 300-yard-wide strip of trees, shrubs, and grasses grew to the edge of the formidable river. A couple of old men sat fishing along the shore, lunch pails beside them: an idyllic country scene except that the patch of shade they sat in was provided by the bow of an enormous tanker tied up along the shore. Trains of barges and oceangoing cargo ships passed before them with highway-like regularity. A vast refinery complex spewed fire and gray gases into the sky on the opposite bank. Tinker-toy pipes stretched from the docked tanker, across the scrubby forest, over the levee and river road and the adjacent town of Taft to another enormous petrochemical refinery.

The refinery complex sprawled over the land, stretching its ducts, pipes, and other tendrils over the road and town like some sort of industrial spider. Flames poured from the tops of high stacks. A labyrinth of conduits, pumps and spigots, generators and holding tanks, hummed, clanged, and belched as they digested crude into gasoline and lubricants. The refinery's security fence abutted the village cemetery. Signs forbade trespassing. The air tasted oily.

Around the next corner stood an imposing nuclear power plant on which high-voltage transmission towers converged. Somewhere out under the river, intake ducts were sucking brown water under the levee and into the reactor building to cool the sun-hot core before returning it to the river. The plant and the adjacent refinery dwarfed even the levees protecting them from the potential fury of the river.

Then farmland again. I passed a couple of small, dirt-poor towns where all the people I saw were black. Then there was a thick grove of magnolia trees surrounding an antebellum plantation house with high columns and shaded porches and expensive cars in the driveway, a white woman in a sunhat reading at a lawn table. Just beyond the magnolias stood some weather-beaten wooden slave quarters and a cluster of trailer homes, black children playing in the yard. Then another petrochemical plant taking up a field-sized plot of land. Another old plantation house surrounded by high fences. In a field of grazing livestock, a dozen weather-beaten houses with rusting roofs, surrounded by chain-link fences. A tall blue sign announced the future site of a Formosa Plastics Corporation complex. A sugar cane field, another plantation house, a row of trailers, a vast chemical plant, a downwind stretch of eye-watering air, another stretch of docked oceangoing ships on the other side of the levee. Tiny rural black hamlets are interspersed with sprawling plantation homes and corporate industrial complexes on both sides of the river from New Orleans to Baton Rouge.

People once referred to these river parishes as the American Ruhr. Now this area is more often called "Cancer Alley." This 90-mile stretch of river is home to 136 petrochemical plants and seven oil refineries, as well as paper mills and power plants. Servicing them all is a nearly continuous chain of port facilities that stretch on both sides of the river like an elongated harbor. River Road is home to a "who's who" of global industrial giants: Exxon, Shell, Chevron, and Marathon, Boise-Cascade and Georgia Pacific, AlliedSignal, Dow, Dupont, Monsanto, BASF, IMC-Agrico, Uniroyal, and Union Carbide.

Together these facilities produce close to a billion pounds of toxins a year, according to the Louisiana Department of Environmental Quality. A substantial portion is released into the environment, making Louisiana the second largest releaser of toxic pollutants in the nation (Texas is first). Sixty-five percent of the 175 million tons of toxic substances released in Louisiana each year come from the ten river parishes (the Louisiana equivalent of counties). Heavy metals and aromatic hydrocarbons, industrial acids and petroleum residues, chloroform and cyanide, formaldehyde and industrial nitrates—all make their way into the air, soil, and water.

A great deal of pollution ends up in the water. Some is released directly into the Mississippi and its associated lakes and bayous. In fact, Louisiana leads the nation in toxic wastes released into surface waters, with 28 million pounds annually. Another 55 million pounds is injected into underground containment pits, where some escapes into aquifers and the rest of the watershed. There are accidents on tankers and pipelines, resulting in oil spills that turn the river black and noxious, contaminating its plants and animals and killing eggs and ecologically important microbes.

It all flows down into the wetlands of the delta and on to the Gulf of Mexico, a toxic deluge from a thousand factories, power plants, and paper mills far up the mighty Mississippi and its tributaries, from Cancer Alley and the petrochemical plants of Lake Charles, Louisiana. Carcinogens and mercury, agents like dioxin that destroy the body's immune system, chemicals that cause birth defects and childhood retardation, others that concentrate in the flesh of fish and birds and then on to the humans that eat them—substances that mix and interact in unpredictable ways once released together into lakes and rivers.

This toxic gumbo is but one component of North America's inadvertent ecological assault on its greatest river, its second largest estuary, and its most productive fishery, the Gulf of Mexico. For generations the Mississippi has been treated as an enormous open sewer draining the refuse of the vast center of the continent. Industrial waste, agricultural drainage, and human sewers all flow into the river and are carried on to the delta and the Gulf.

This is the story of America's homegrown Black Sea crisis. The elements will be familiar to you: a polluted river, a neglected sea, and a vast and disparate array of culprits living in far-flung legal jurisdictions. But unlike the Black Sea, there's still time to do something about the gathering crisis in the Gulf, and there's a wealthy country to provide the means.

It all begins with a Dead Zone.

❧

EVERY SO OFTEN SOMETHING STRANGE HAPPENS on a stretch of Gulf coast shoreline. Fish, crabs, and shrimp all but throw themselves into the arms, baskets, and hand nets of people wading in the beach surf. Thousands of redfish and drums, gulf shrimp and blue crabs surrender themselves to celebrating townspeople in Louisiana, Upper Texas, and Alabama as if bidden ashore by a benevolent god. Baskets are effortlessly filled with fresh seafood. In a few hours a single person can collect a hundred pounds of shrimp. It's a miraculous event that brings together entire communities in a celebration of the sea's seemingly wondrous bounty.

Gulf folks call it a "jubilee." The reality, at least for the sea's creatures, is less than jubilant. They aren't presenting themselves as gifts to man but trying desperately to escape suffocation.

Out in the Gulf of Mexico lies the Dead Zone, a 7,000-square-mile swath of oxygen-less water in which few creatures can survive. Fish, crabs, and shrimp must shun this biological void or perish. Things that can't leave die. Slow-moving starfish, urchins and snails, immobile mollusks and anemones, and burrowing worms are snuffed out in untold millions.

Scientists on the Black Sea would recognize this Dead Zone, a seasonal version of the permanent blanket of death that appeared off Odessa and ultimately laid waste to the entire Black Sea basin. In Romania and Georgia, Turkey and Ukraine, they know this is an ominous portent to be taken seriously: seasonal hypoxia—the scientific name for the lack of oxygen in sea water—has a way of becoming a permanent condition.

The Dead Zone forms every spring and summer, following the prevailing currents from the mouth of the Mississippi down the Louisiana coast and on into Texas. It's been likened to stretching Saran Wrap over an area the size of New Jersey and suffocating anything that cannot escape.

Because it usually stays several miles offshore, the Dead Zone's effects are mostly out of people's sights and minds. Shrimp fishermen know it's there; they have to spend time and fuel to travel beyond it before lowering their nets. Scientists who study it know how devastating it is to marine ecology. But most people never see its effects first-hand.

Except in those rare times when it comes ashore, driving all animal life before it. Those that don't leap into the baskets of people on the beach perish in the undersea vacuum.

"Jubilees" have been a fixture of life in Alabama's Mobile Bay since 1867, the result of oxygen-poor deep water being forced by weather and tides into shallow coastal areas. In Alabama, locals told me the number of jubilees has remained fairly steady as far back as they can remember, an annual occurrence somewhere along the bay. That jibes with oxygen readings taken in Mobile Bay for the past twenty-five years: They've remained more or less stable.

The area around the Mississippi delta is another story. The Dead Zone off the Louisiana coast became a perennial phenomenon in the 1950s and grew dramatically in size and seriousness during the early 1990s. In the summers of 1990 and 1996 residents of the bayou fishing port of Grand Isle were treated to massive jubilees. Residents and visiting tourists scooped up hundreds of pounds of fish and shrimp in hand nets, filling ice chests and bushel baskets. Most fish wound up dead on the beaches, cast aside by beachcombers since they were small, the wrong species, or already decaying. Thousands of sea birds wandered the sandy shores feasting on menhaden, sardines, and juvenile redfish.

In western Louisiana and upper Texas, fish kills have lately become an annual spring event. Hundreds of thousands of dead hardhead sea catfish cover the sea in putrid sheets. Others blanket beaches, puncturing the feet and even automobile tires of beachgoers with their sharp, spiny fins. In 1994 more than half a million fish are estimated to have perished on this stretch of coast. The Dead Zone and the Mississippi are thought to be to blame.

In 1992, Hurricane Andrew just missed New Orleans, a densely populated city that lies mostly below sea level and is almost impossible to evacuate owing to the levees and swamps that surround it. Instead Andrew smacked into a sparsely populated part of Louisiana bayou country, a hundred-mile-wide skirt of marsh and swamp running from New Orleans to the Texas border. The storm surge buried a large swath of the barrier coast under fifteen feet of hypoxic water. Millions of fish died along the Louisiana coast. From the air observers saw a reef of dead fish 5 to 10 feet wide and 20 miles long. The storm surge also pushed into the marshes and bayous, where an estimated 200 million fish died in the following weeks from the decomposition of leaves and untold tons of organic sediments mixed up in the swamps and bayous by Andrew's powerful winds. Ninety percent of the fish in the Atchafalaya River's basin were killed, and untold numbers of birds and marsh animals.

The Louisiana shore had received a taste of what was occurring in the deeper waters of the Gulf.

∞

FROM THE OBSERVATORY TOWER of the Louisiana Universities Marine Consortium (LUMCON) building you can see the crushed remains of several dwellings scattered through the tall marsh grasses surrounding the fishing village of Cocodrie. One structure appears to have been a trailer, its metal sheeting rusting off its skeleton. Another was once a wooden frame house but is now a sodden pile of lumber. A third is so damaged not even the mind's eye can put it back together. Hurricane Andrew tore all of these from their moorings, flung them over the road, and broke them up in the marsh.

The LUMCON building itself was built to be hurricane proof, a modern concrete structure resembling a small airport terminal. When Andrew hit it it had thick, wind resistant glass, a "disposable" ground floor designed to allow a 10-foot storm surge to flow under the building, and bunker-like concrete-and-steel pillars supporting walls and ceiling. Unfortunately it also had one of those flat roofs that are covered in large pebbles to assist with rainwater drainage. The 145-mile-per-hour winds picked up these pebbles and strafed the building with them like a machine gun. "Hurricane-proof" windows were destroyed along with much of the building's contents. When flood waters retreated, the ground floor was buried in several inches of mud.

All of Cocodrie's homes are now on stilts. New trailer homes are suspended 16 feet off the ground from stout wooden frames, some with wheels blowing in the wind. Rows of houses perch above two stories' worth of open air in an effort to avoid the next storm surge. So vulnerable is this town—located on the marshy fringe of the Gulf of Mexico—that the 16-foot stilts are required under new structures before electricity and water lines can be connected. That is since the marshes that once protected Cocodrie from storm surges are vanishing under the waves.

In Cocodrie I met with the scientist most knowledgeable about the Dead Zone, Dr. Nancy Rabalais, a Texas native with a hint of a drawl who has been studying this massive hypoxic zone since shortly after coming to LUMCON in 1983. Since then she's devoted considerable time and effort to making people aware of the scale and seriousness of the Dead Zone. She's testified before Congress, written backgrounders for state officials, and met more journalists than she can count. "I need to take the science to the public and that's thrown me into the policy arena, whether I like it or not," she says.

The Gulf of Mexico is polluted by many things: pesticides and oil, toxic releases, and household garbage. But as in the Black Sea, the most severe degradation comes from nutrient pollution. The Mississippi, like the Danube, drains half a continent. The watersheds from part or all of thirty-one states and two Canadian provinces ultimately flow into it. This vast region comprises 42 percent of the lower forty-eight American states and includes almost the entire breadbasket in the Midwest and Upper Plains. From the cornfields of Iowa and Indiana to the livestock paddocks of Montana, all manure and fertilizer run-off eventually winds up in the Mississippi. So do the waters from sewers and storm drains from Denver and Minneapolis to St. Louis and New Orleans. It all empties into the Gulf.

Between 1955 and 1980, nitrogen fertilizer use in America increased by 600 percent as artificial chemical fertilizers became cheaper and more widely available. In the Mississippi River, concentrations of dissolved nitrogen and phosphorus doubled. It's the increase in nitrogen that's believed to be behind the formation of the Dead Zone in the Gulf.

Nobody was monitoring nutrient and dissolved oxygen levels off the Louisiana shore until the 1970s, when oceanographic surveys were undertaken as part of offshore oil and gas exploration. So Rabalais and her colleagues have looked under the sea floor for information about the past.

The Mississippi is aptly nicknamed "Big Muddy." It carries an enormous load of sediments as it drains the croplands and mountain valleys of much of America. When the water enters the Gulf of Mexico it slows down and the sediments fall out onto the ocean floor in a great plume. So much sediment is deposited that it forms thick annual layers on the seafloor. To determine the age and causes of the Dead Zone, Rabalais and other scientists have taken deep sediment cores of this part of the ocean floor. Once analyzed, they reveal several centuries of Mississippi River basin history, an account written in the piled remains of billions of microscopic plants and animals.

Like the rings of a tree, the quantity of diatom shells in each layer indicated what sort of growth conditions the algae had experienced. Cores dating as far back as the 1700s show low, stable quantities of algae. There was a peak of productivity in the mid-1800s, when American settlers cleared the Midwest for crop planting, triggering an initial wave of soil erosion. A major flood in 1927 brought huge quantities of soil and nutrients, triggering another temporary peak. Then things remained stable again until the mid-1950s, when large algae blooms began forming on a regular basis.

The remains of two other creatures fill in more of the story. *Ammonia parkinsoniana* is a shelled protozoan that does well in hypoxic conditions. *Elphidium*, a similar microscopic animal, does poorly. The relative quantities of the two species began changing rapidly in the mid-1950s, with *Ammonia* completely dominating the assemblage by the mid-1970s. *Elphidium* and other hypoxia-intolerant species suffered population crashes. Starting in the mid-1950s, sediment layers started to contain increasing quantities of glauconite, a clay formed under low oxygen conditions. Scientists were assembling compelling evidence that the Dead Zone was fueled in large part by chemical fertilizer run-off, allowing it to spread ever larger off the Louisiana coast.

When first mapped in 1985, the Zone covered 3,500 square miles of the Gulf, extending from the seafloor (where most life is concentrated) to include 10 to 80 percent of the total depth of the water. It grew slowly until the Midwest's 1993 flood, which drained huge quantities of nutrients into the river and Gulf. That summer it doubled in size and has stayed that way ever since: 7,000 square miles in 1994–1997, with a slight reduction in 1998 that Rabalais thinks may be due to wind and weather rather than an improvement in oxygen levels. The problem, it appears, has become chronic.

She showed me a video of the Dead Zone seafloor shot from a remote-controlled underwater camera. No fish could be seen at all. Occasionally there are a few stray mantis shrimp—swimming if dissolved oxygen readings are above 1.5 mg per liter, strewn dead on the bottom when readings fall below that figure. Then you begin to see dead snails, anemones, and bottom creatures. Burrowing worms poke out from the sediment, desperately reaching for oxygen. Sea stars stretch their limbs up into the water in a vain effort to escape. Sulfur-oxidizing bacteria rise up from within the sediments, forming ugly white mats that hover above the seafloor.

In the fall the area begins a seasonal recovery. The Mississippi's flow is reduced, bringing fewer nutrients, while hurricanes and other storms stir and mix the water, breaking up the hypoxic Zone. Some animals return for a few months—almost exclusively those species whose eggs and larvae float around in the currents before maturing. Creatures whose eggs hatch where they are laid just aren't there and haven't been for more than a decade. With less life there year-round, there's less food for predators and fewer shrimp and crabs for fishermen to gather. What community is left behind is simpler, weaker, less able to respond to an outside threat: an oil spill or an alien comb jelly, toxic poisons or destructive fishing practices.

"Nobody can say for sure what will happen," says Rabalais. "What's certain is that we'd have a healthier system if the nutrient load in the river went down."

But that's a very difficult thing to accomplish, involving, as it does, most of the farmbelt, hundreds of city wastewater plants, and, strangely enough, a group of generals waging war on the angry river.

❧

PRIMGHAR IS OVER A THOUSAND MILES from the Gulf of Mexico, but it feels even farther away. It's a small farming town in north-western Iowa, a community clustered not around a river or stream but on a leafy courthouse square. In every direction fields of corn, wheat, and soybean roll gently to the horizon. It seems they must go on forever. My great grandfather raised chickens and hogs here in the center of a vast continent, as far from the ocean as one can get. Perhaps he took his cue from his father-in-law, my great great grandfather, a second generation prairie farmer at a time when prairie farming was still young. "What's the use of coastal places?" my grandfather often heard him say. "You can't grow corn there worth a darn."

Newfoundland was built on cod, but in Iowa corn is king. Today the state grows 21 percent of the nation's corn, more than any other state. Most of it is used as animal feed, which is why Iowa also leads the nation in pork production. A full 25 percent of American pork comes from Iowa farms, which produce 20 million corn-fed hogs every year. Sixteen percent of American beef cattle are raised in the state, and Iowa grain accounts for almost a fifth of American grain exports. The state department of agriculture's motto is more accurate than most: "If farmers are prosperous, all Iowans will prosper."

But when the first settlers came to the state in the early 1830s, they found much of the country useless for agriculture. In 1860, set-tlers regarded northwest Iowa as a "wilderness of ponds and sloughs, alternating with patches of tough prairie grass, vegetation of such rank growth as to be useless for hay or pasturage, fit only for the habitat of muskrats, coyotes, frogs and mosquitoes." Stagnant ponds and so-called wet prairies dominated much of the landscape not only in Iowa but throughout large portions of the present-day cornbelt in Illinois and Indiana. In spring the flat, slow-draining landscape became as soaked as a wet sponge and stayed that way far into the growing season.

Two technical innovations helped transform these "worthless" lands into the "breadbasket of America," a region that now provides much of the world's agricultural surplus. Both would have dramatic effects on run-off into the Mississippi and Gulf of Mexico.

The first was tile drainage, the elaborate reengineering of the soil structure in Iowa, southern Illinois, and Indiana. In the 1870s and 1880s farmers found an ingenious way to drain these marshy lands. A layer of tiles was inserted a few feet beneath the land's surface and placed in such a way that they intercepted groundwater and channeled it into artificial drainage ditches. These in turn delivered the "short-circuited" groundwater into rivers and streams leading to the Mississippi. Once drained, the land proved to be the richest in the country. The technology spread rapidly. By the early 1880s, 1,140 factories were churning out drainage tiles; Indiana had over 30,000 miles of tile drains and Ohio had 20,000 miles of public ditches. Hundreds of millions of acres of highly productive land came into cultivation, fueling further expansion of the railroads and the nation. It is around this time that Nancy Rabalais's core samples indicate the first wave of excess nutrients in the faraway Gulf.

Today about 50 million acres of the Mississippi basin have been drained with tiles and ditches in an effort to improve farming.

The second innovation came after World War II with the advent of chemical fertilizers. It was discovered that nitrogen fertilizers could be produced on an industrial scale from crude petroleum, dramatically reducing the cost of fertilization to the farmer. To produce a good crop, cornfields need more nutrients than even Iowa's fertile soil can provide; 1.2 pounds of nitrogen per bushel is the rule of thumb. Cheap fertilizers allowed Iowa, Illinois, and Indiana corn farmers to become incredibly productive. Fertilizer use increased by 950 percent between 1950 and 1970, then doubled again between 1970 and 1980, a seventeen-fold increase. Many fertilizer plants were established in Louisiana's chemical corridor because of its easy access to the state's oil reserves and Midwestern farms via Mississippi barges. Under the barges an incredible quantity of wasted nutrients rushed towards the Gulf.

After the Dead Zone was discovered, scientists worked to trace the origin of the nutrients spilling into the Gulf. Not surprisingly, more than 90 percent came from the Mississippi and its distributary, the Atchafalaya. What was surprising to many is that most of the nutrients escaping the Mississippi's mouth originated in the Upper Midwest, hundreds of miles upstream. United States Geological Survey (USGS) studies found that more than 70 percent of the total nitrogen delivered to the Gulf by the Mississippi originates north of the Ohio River confluence. Iowa alone is responsible for about a quarter of the million tons of nitrate-nitrogen entering the Gulf each year. USGS also estimated that 90 percent of the nitrogen pouring out of the Mississippi's mouth came from so-called non-point sources, of which agricultural run-off is the most important. Municipal sewers, industrial effluents, and other point sources accounted for only about 10 percent of this nitrogen load. The Corn Belt, in other words, is by far the largest cause of the Dead Zone.

Tile drainage, as it turns out, delivers large quantities of nutrients into rivers and streams. In a study of a watershed in north central Iowa, James L. Baker and H. P. Johnson of Iowa State University compared nitrate concentrations in run-off from pastures, conventional row-cropped corn and soybean fields, and similar fields with tile drainage. The two types of crop fields received the same levels of fertilization. Tile drainage fields had average nitrate concentrations six times higher than that of ordinary fields and twelve times that of pastures (which received only manure from the livestock grazing upon them). Tile-drained fields lose 15 to 20 pounds per acre of nitrate nitrogen in this fashion. James Baker says that nitrate concentrations of 10 to 30 mg per liter are not uncommon in the drains—that's one to three times the national safe drinking water standard.

Once in surface waters, these nitrates have effects both near and far. Tile drains run into ditches and on to rivers. In Iowa, the Raccoon River alone captures run-off from 6,263 farms covering 2.3

million acres of land; the Raccoon also provides drinking water for 270,000 people in central Iowa, including the city of Des Moines. The river is laden with nitrates, often at levels 20 to 40 percent higher than federal standards. High levels of nitrates can cause the "blue baby" syndrome in which infants are robbed of the ability to deliver oxygen in their blood. There's also an apparent link between high nitrate levels and bladder cancer in women over fifty-five. Studies in Iowa found that female bladder cancer rates were higher where nitrate levels in drinking water were also high. To avoid these public health problems, Des Moines spent $4 million building the world's largest ion exchange system to remove nitrates at the city's water purification plant. Smaller communities lack these resources. An estimated 175,000 Iowans drink from water systems and private wells with nitrate levels exceeding federal standards.

But most of these nutrients are carried far from Iowa. Most nutrients eventually wind up in the Mississippi River, the ultimate drainage ditch for America's breadbasket, and begin their long journey towards the Gulf.

☕

ONE OF THE STATUES STANDING on New Orleans's Jackson Square turned out to be a young woman. She had covered her skin and the long dress she was wearing with metallic silver spray paint and mounted herself on a wooden base of the same color. She stood absolutely motionless for hours, except to perform a very mechanical curtsy whenever a passer-by dropped coins into her collection box. Two very small children tap danced on an adjacent corner, hoping to attract tips from passing throngs of drunk, sunburned tourists; a six-piece jazz band on the other side of the square was playing "When the Saints Come Marching In," luring more tourists off the never-ending party on Bourbon Street. A typical late morning in the French Quarter, where the bourgeois and burlesque are retailed side by side to a vacationing Middle America.

Two centuries ago Jackson Square opened onto the Mississippi waterfront. Today you can't get so much as a glimpse of the river.

The huge artificial hillside of a three-story-high levee blocks the view. Only from the top of the levee can the mighty river be seen, a fast-moving concourse crowded with oceangoing tankers and tugs pushing long trains of heavily laden barges. Wakes from passing container ships throw brown waves against the levee's concrete façade. From here, New Orleans's great illusion is broken: the fiction that this metropolis sits on dry, solid ground beside the world's third largest river. You can turn around and look down on Jackson Square, the teeming French Quarter, the tall office buildings along Canal Street and see that it all sits *below* the brown surface of the Mississippi. Far below in fact. Much of this city of 1.2 million lies 10 or 15 feet under the level of the river; only the massive levees and enormous pumping stations keep the river from flooding the metropolis. United States taxpayers pay tens of millions each year to allow New Orleans to continue to thrive on land that is not only lower than the river but below sea level.

France chose a poor site for a major city. No matter that the last couple hundred miles of the river consisted of low-lying marshes and swamps built on the river's own silty sediment loads. Paris wanted a port at the mouth of the Mississippi to command shipping in the vast North American interior. In 1699 a small "island" of firm ground was located and a new city was established there a few years later.

From the outset, New Orleans was at the mercy of the Mississippi's spring floods. The first skirmishes with the river began in 1717, when Sieur de Bienville, the founder of New Orleans, ordered the design and construction of a 3-foot tall, mile-long dike to protect the new settlement. As more people came to the city and surrounding farms, levees continued to expand in height and length. So began a vicious spiral that continues to this day: New levees allowed new developments on the Mississippi's flood plain which, in turn, increased the political and economic pressure to build bigger, longer levees to protect increasingly valuable property. By 1850 the river was almost completely leveed from New Orleans to Baton Rouge on the east bank and to the mouth of the Arkansas River on the west.

Few considered the implications of cutting the enormous river off from its floodplain. This radical reorganization of the Mississippi Valley's plumbing system began a process that not only worsened flooding but would eventually cause large parts of southern Louisiana to vanish into the sea. It would allow nutrients from the cornbelt to speed directly to the Gulf, while undermining the marshes and estuaries on which so much Gulf marine life depends. Two centuries of short-term thinking have created ecological problems of colossal proportions.

Once early levees were built, planters discovered that the dark soil of the Lower Mississippi flood plains was ideal for the production of sugar and, more important, cotton. All that was needed were strong levees to defend against floods and a large labor supply to plant and harvest the fields. Tens of thousands of slaves were brought in from other states to clear and work the land. Their owners made huge profits and were able to finance the construction of opulent plantation homes along what is now River Road in Louisiana and throughout the Yazoo River Delta in Mississippi (popularly known as the Mississippi Delta). They also financed the construction of ever-larger levees to protect their valuable holdings. The federal government gave vast tracts of "worthless" swamp land to the states, which used the proceeds from their sale to finance levee construction. These floodplain swamps, which had served as natural flood control reservoirs, were now sealed off from the river behind levees and cleared and drained to become valuable farmland that, in turn, had to be defended from flooding.

Confined to the main channel and no longer able to spill its banks, the river could only flood upward. Flood crests in the main channel grew higher as more and more of the floodplain was sealed behind levees. Massive floods in 1858 and 1859 topped or knocked down many levees. In response, the federal government continued building ever-larger levees.

In 1879, coordinated flood control was turned over to the U.S. Army Corps of Engineers. For much of the nineteenth century, West Point was the only American institution that trained civil engineers, and the military dominated the profession. The Corps declared war on the river and expended considerable effort to domesticate it. Unfortunately, it did so using faulty hydrologic theory.

The Corps assumed that once confined to its main channel, the river would scour and deepen its own bed. It was thought that the river would thus increase its capacity more than enough to make up for its greater volume. Escape valves—reservoirs, spillways, access to flood plains and distributary rivers—were considered counterproductive. The Corps built levees to seal off distributary channels from the slow-moving bayous of southern Louisiana to the massive Cypress Creek, which in times of flood had a flow exceeding that of the flooding Danube. More and more water was channeled in between the levees and rushed towards the outlet in the Gulf.

The more the Corps engineers practiced flood control, the worse the floods grew. Levee-busting floods came in 1882, 1884, 1890, 1897, 1903, 1912, 1913, and 1922. Each time the Corps responded by extending and raising levees, which only increased the flood crests. A 7.5-foot levee protected Morganza, Louisiana, against the flood of 1850; by the mid-1920s it stood 38 feet tall. The river's scouring was not making up for the loss of thousands of square miles of Lower Valley flood plains. The final proof came in 1927 when a massive flood breached forty-two levees and flooded seventeen million acres in seven states. Three hundred died and 637,000 were made homeless. To save New Orleans the Corps dynamited an upstream levee, flooding poor, rural south Louisiana parishes instead. At the height of the flood the Mississippi's flow nearly tripled, to 2.4 million cubic feet of water per second.

After the 1927 flood the Corps reassessed its strategy. The new plan called for three emergency spillways to divert river water during a major flood, relieving stress on levees and protecting New Orleans. Thirty miles above the city the Corps has on several occasions diverted flood waters into one such spillway leading directly to Lake Pontchartrain. Another spillway upriver from Baton Rouge has been opened only once. That was during the flood of 1973, when the Corps very nearly lost its war with the river.

THE MISSISSIPPI'S REAL DELTA begins three hundred miles inland from its current mouth, in north central Louisiana. This is the point where the river stops collecting water from tributary rivers and starts releasing it through a network of distributary waterways spreading out over thousands of square miles of southern Louisiana. Its main channel changes course every thousand years or so, splaying water and silt across the delta like a loose hose on a driveway. Each time it changes course, the silt it deposits begins creating new land around its mouth, a new delta. As a delta grows and matures, the river's path through it becomes more and more circuitous and choked by sediments. Water flows less freely, increasing pressure upstream, which the river relieves by shunting more of its flow down a distributary. That distributary channel eventually "steals" the river and begins building a new delta with the hundreds of mil-lions of tons of sediment delivered by the Mississippi each year. The old delta, now deprived of regular silt supplies, begins to slowly sink away. The Mississippi's current channel, the one that carries it past Baton Rouge and New Orleans, is only one thousand years old. Prior to that it flowed down what is now slow-moving Bayou Lafourche to enter the Gulf at a point 60 miles west of its current mouth. Many of Louisiana's bayous—slack, brackish waterways that resemble a cross between a river and a lake—were once the raging main channel of the Mississippi. Virtually all the land in southern Louisiana was created and sustained by this process, which slowly built the bayou country from sediments the river brought here from the Upper Plains, the Midwest, and the slopes of the Rocky Mountains.

In its effort to control the river, the Corps has wreaked havoc with this process. As a result much of southern Louisiana is literally van-ishing beneath the surface of the Gulf of Mexico. And the Corps finds itself in a never-ending struggle to stop the river from making one of those jumps to a new channel.

Once the spillways were in place, the Corps sealed the Mississippi off from all distributary rivers save one. The Atchafalaya River in Louisiana already carried over 20 percent of the Mississippi's flow, and the Corps intended to use it as the last defense in case of a massive flood. But during the 1950s the Corps realized that more and more water was pouring into the Atchafalaya every year. It already carried 30 percent of the flow but in the near future it might reach 70 percent. The reason: During the 1830s the state of Louisiana had dynamited an ancient log jam blocking the Atchafalaya, while a private entrepreneur straightened the Mississippi near its confluence with the Atchafalaya. For more than a century, water had been flowing down this once minor distributary, scouring its channel and thus increasing its flow. The newly opened distributary had a channel 35 feet lower than the Mississippi's main channel; it was now a far straighter, shorter, faster route to the Gulf. Once again the river was changing its course in accordance with the imperatives of gravity and geology.

The economic consequences of allowing this jump were deemed too great. If the Mississippi no longer ran past Baton Rouge, the billion dollar industries of Cancer Alley, and the great port of New Orleans, they would all wither like unwatered plants. Instead the river would enter the Gulf at Morgan City, 140 miles west of New Orleans. The newly extinct main channel would turn into a saltwater estuary, a long arm of the sea, a backwater bayou. In 1954, Congress passed a law forbidding the Mississippi from making the "Big Jump." The Corps was ordered to stop the Atchafalaya from taking any more than 30 percent of the Mississippi's flow.

The result was the Old River Control Structure, 500,000 tons of concrete and steel blockading the Mississippi's path into the Atchafalaya. Completed in 1963, Old River was designed to maintain control during a massive flood of 3 million cubic feet per second, considerably larger than the 1927 flood.

The river defeated the design. During a severe 1973 flood the river tore away one of the structure's guide walls, scoured a 55-foothole underneath it, and shook the entire structure like a child's toy. Engineers were only able to save it by dumping 250,000 tons of boulders and debris into the hole before the river could carry the entire structure away and charge into its new channel. The 1973 flood had a flow of 2.23 million cubic feet per second, well under Old River's design capacity. Since then a new control structure has been built along with a large hydroelectric dam. The Corps says that with all three structures in operation the river can be held as long as Congress bequeaths the funds to operate and maintain them.

∞

AFTER THE 1973 FLOOD, SCIENTISTS NOTICED something miraculous happening in the lower Atchafalaya. New islands had appeared near the Atchafalaya's mouth and they grew with each subsequent spring flood. Shoals of mud and sand appeared in what had been deep water. Marsh plants began colonizing the area, drawing large numbers of migratory birds. The birds were soon joined by crowds of migrating scientists coming to study what is an extremely rare event in human terms: the formation of a new river delta.

Despite artificial constraints on its flow, the Atchafalaya is steadily building marshy delta land. At its mouth in the Atchafalaya Bay more than 16,000 acres of new land have appeared. Deltaic islands are appearing where Atchafalaya Bay meets the sea. Nature is preparing for the Mississippi's inevitable shift.

The Atchafalaya delta's growth stands in sharp contrast to the condition of the rest of southern Louisiana's bayou country. The Corps closed the Mississippi's other distributary channels decades ago. To protect New Orleans and the "American Ruhr" the Corps built high levees all the way down the main channel, forcing the river straight to the Gulf of Mexico. There it drops its sediment load, laden with nutrients and toxins, into 120-foot-deep water. Deprived of these sediments, Louisiana's bayou country is literally disappearing under the sea.

The marshes that make up virtually all of southern Louisiana are vanishing at the staggering rate of 25 to 35 square miles per year. Since 1930, Louisiana has lost more than a million acres—an area larger than the state of Rhode Island—as marsh and swamp have turned into open ocean and lakes. In human memory, entire barrier islands have gone, vast marshes have turned into open bays, and once protected towns have become exposed to the full fury of the Gulf's hurricanes. If the loss continues, nineteen town sites will be under water by 2050 along with most of the southern part of the state.

The levees the Corps built to protect New Orleans from the river have ironically facilitated the destruction of the city by the sea. Levees cut the marshes off from the river that sustained them with sediments and fresh water. With freshwater flow much reduced, salt water from the Gulf began intruding, killing marsh plants that could not tolerate heightened salinity. Canals cut through the marshes by the oil and gas industry compounded the problem. Without new sediments to nourish them, other marshes sank away under their own weight.

By 2050 the marshes protecting New Orleans will be nearly gone. The city will lie on an island directly abutting the Gulf, a sitting duck for a powerful hurricane. It's estimated that a hurricane's storm surge loses 3 inches of height for every mile of marsh it passes across. Between 1930 and 2050, the equivalent of more than 30 miles of marsh will have vanished between the city and the gulf. A direct hit by a hurricane the likes of Betsy (1965), Juan (1985), or Andrew (1992) could overwhelm protection levees and put the French Quarter under 20 feet of water. A Louisiana State University storm surge model showed that a hurricane could push an 18-foot storm surge into Lake Pontchartrain, a surge topped by 10-foot waves that would easily overwhelm newly completed levees protecting the downtown. This is a nightmare for evacuation planners since there are few escape routes over the levees encircling the city. Planners call for "vertical evacuation" in which hundreds of thousands take refuge in the Superdome and the stairwells of downtown high-rise office buildings. Many doubt the efficacy of this type of evacuation.

The roads and highways of south Louisiana are posted with Hurricane Evacuation Route signs with arrows pointing motorists towards firmer ground 50 or 100 miles north. Against their advice I drove south from New Orleans through stands of cypress swamp and marshland. Although it was January and there had been little rain, a mile-long section of Route 90 was sandbagged, and yellow signs warned "Road Floods in Storms." This was 50 miles inland from the Gulf. A single strip of asphalt follows Bayou Petit Caillou more than 40 miles from Houma to Cocodrie. The road follows the only high ground there is around here: a natural levee 4 feet above the sluggish bayou. A chain of towns is perched along the curb. Many homeowners have a pick-up truck parked on the shoulder in front of their house and a large shrimp trawler tied up to the bank of the bayou in the backyard. Lawn furniture, children's toys, garden hose, propane grill, and 60-foot, steel-hulled offshore fishing vessel: all the accouterments of Terrebonne Parish life. Many people commute to work not on the road but on the bayou. After a day of work on the Gulf they offload their catch and tie up in their own backyard. In the evening their fishing vessels are parked like automobiles on a suburban street.

Cocodrie is the end of the road, surrounded by marshes on all sides and only a mile from the open Gulf. With all those houses and trailers perched on two-story-tall stilts, Cocodrie looks like a giant's collection of house-shaped bird feeders. When hurricanes come this way, Cocodrie must evacuate early: It is 100 miles of tertiary roads to the nearest Red Cross shelter. The Red Cross won't evacuate anybody to locations south of Interstate 10 since the land is too low-lying to be considered safe. More than 2 million people live in the evacuation zone. But the risk is far greater for those living at the end of the road, in places like Cocodrie and the resort town of Grand Isle, the offshore oil towns of Morgan City and Port Fourchon, and the boat-building town of Galliano (whose residents, coincidentally, built the ship that took me to Antarctica). Not only are they closer to the Gulf and farther from safety, but the intervening roads are also vulnerable to flooding. By 2050 so much surrounding land may have been lost that the towns will have to be abandoned. If Cocodrie's residents aren't driven out by flooding and lost roads, they may leave since they can no longer find shrimp for their trawls.

The loss of the Louisiana bayous is not only a human disaster but an ecological crisis affecting the entire Gulf of Mexico. The bayou country is the nursery for much of the Gulf's marine life. It is a vast estuary, a sheltered place where fresh and salt water meet and mingle. Here amid the reeds, marshes, and floating mats of vegetation, huge numbers of juvenile shrimp and fish feed on the ample supply of fallen plant bits and little creatures living on the roots and leaves of living plants. The network of shallow channels between marshy plants protects juveniles from predators. Many species evolved in conjunction with such estuaries; juvenile shrimp, for instance, require habitats with different salinities at different stages of growth, habitats that exist only in the Gulf's estuarine marshes. When they are large enough, they migrate offshore and, later, lay eggs that currents will carry back towards the marshes. Other species such as oysters can live only in the brackish water of the estuaries.

As to the economic value of the marshes, a popular Louisiana bumper sticker says it well: "No wetlands, no shrimp." Not only shrimp, but 98 percent of commercially important fish species in the Gulf depend on these wetland nurseries. These species support a billion-dollar Gulf coast fishing industry that accounts for 40 percent of the nation's total fish catch. Louisiana alone harvests 25 to 35 percent of the nation's commercial fish catch. It leads the nation in the harvest of menhaden, crabs, shrimp, and oysters. Recreational fishermen contribute another $944 million to the state's economy. Throughout the Gulf, some two hundred thousand people rely directly on commercial or recreational fishing for their livelihoods. Millions more indirectly rely on these industries. As the estuaries vanish, so do fish and jobs.

Lost wetlands may also make the Dead Zone worse. When they are healthy, wetlands also play an important part in cleansing river water of toxins and excess nutrients before it reaches the sea. Wetland plants assimilate nutrients and some toxic chemicals so efficiently that many communities construct artificial wetlands to help clean municipal wastewater. But instead of running through many miles of cleansing swamps and marshes, the Mississippi and its enormous nutrient load are funneled between the levees directly to the Gulf.

❧

THERE ARE NO EASY WAYS to address the Mississippi's dysfunctional relationship with the bayou and the Gulf.

The Dead Zone is the result of more than a century of agricultural engineering across America's heartland. The problem is worsened by nearly two centuries of Corps projects designed to reengineer the plumbing of the continent's largest watershed. The levees and associated flood control systems undercut the basis of life in the Gulf of Mexico by destroying its most important nursery; ironically, the loss of these wetlands may ultimately destroy the very communities and infrastructure the levees were designed to protect.

But what has been done cannot be easily undone. Tens of millions of people now live in the Mississippi's floodplains. The levees cannot simply be removed since we have built so much behind them. They protect cities and farms, power plants and factories, airports and highways.

America's cornbelt relies on chemical fertilizers and tile drainage to maintain its impressive yields. The cornbelt is the engine of a $30 billion industry, one that has made America by far the largest grain exporter in the world. As the world's population increases and the quantity of cropland in China and other developing nations decreases, more and more of the world's population will rely on Mississippi Valley farmers for their food supply. Measures to protect the Gulf that significantly reduce crop yields will prove impractical on political, economic, and social grounds.

Solutions will therefore be complicated and expensive. They include "big fix" engineering projects to reconnect the river to the deltas it created, politically difficult legislative changes to land use and pollution practices, and tens of thousands of small initiatives undertaken by individual farmers, property owners, and municipalities to reduce their impact on the watershed while improving the efficiency of their operations.

∞

In Baton Rouge I met with an official from the governor's office to talk about Louisiana's multi-billion dollar plan to stop the southern third of the state from falling into the sea. I wasn't sure what to expect. Louisiana has a long tradition of "colorful" politics, a euphemism for generations of corruption, cronyism, and pandering to the industrial giants of Cancer Alley. The state isn't known for its environmentalism. Len Bahr, the man I was meeting, had served as a senior advisor to three successive governors. The first, Buddy Roemer, initiated much-needed environmental reforms but his personal behavior was so erratic that most reforms were reversed before he was voted out of office. His successor was three-time Governor Edwin Edwards, a rabid anti-environmentalist who was served federal indictments for corruption and bribe-taking. The current governor, Mike Foster, had waged war on the Tulane Environmental Law Clinic after it helped a poor black community stop Formosa Plastics Corporation from building the world's largest polyvinyl chloride plant in their town. Law clinics in Louisiana are now banned from representing litigants whose incomes are greater than $20,563 for a family of four. What sort of environmental advisor would I be meeting?

I was soon disarmed. Bahr met me for breakfast at his favorite diner. He was a fit, middle-aged man wearing jeans and a baseball cap featuring the Cookie Monster from *Sesame Street*. He had a Ph.D. in ocean-bottom ecology and had worked in the marine sciences department of Louisiana State University before being appointed to the governor's office. He spoke with genuine enthusiasm about bayou geology, wetland preservation, and deltaic hydrology. He had formed a close working relationship with the Coalition to Restore Coastal Louisiana, a conservation group headed by an environmental lawyer. By his own admission, Bahr had sometimes had to keep a low profile to keep his wetlands restoration strategy alive.

The best way to reverse Louisiana's staggering land loss, Bahr has argued for years, is to reconnect the river to the delta country it created. But progress towards that goal has been painfully slow. People have moved into the areas behind the levees and many interests resist major hydrological changes even in sparsely populated bayou country. Getting anything done "is going to require some leadership," he says. "It's going to require breaking some political eggs."

It's also going to require a staggering amount of taxpayer money. After years of consultations with scientists, environmental activists, and government agencies, a joint federal-state task force has released its "big fix" plan to save the southern parishes. It calls for the construction of a series of gates, locks, pumps, dams, and diversion structures along the river and dozens of bayous. Shorelines, river banks, and entire Gulf shore islands would be stabilized. Artificial reefs would be constructed offshore. All these projects share one intention: to restore more natural drainage and salinity patterns to the region.

The price tag: an estimated $14 billion, much of which would have to come from federal coffers.

Getting that kind of money would be tricky for any state. Louisiana is further handicapped by its "colorful" politics, which have lowered the state's credibility in Washington. Bahr argues that the problem is so severe that those billions will have to be spent one way or another. "You either spend it protecting and rebuilding marshes or you spend it rebuilding New Orleans from the next big hurricane," he says. Hong Kong found the money to build a $22 billion airport. Certainly America can find $14 billion to save the homes of two million people?

Federal taxes are already funding $380 million in Mississippi diversion projects. When completed they are expected to prevent about 20 percent of the overall land loss projected for 2050. Bahr is happy that the projects exist, but says they're simply "way out of scale." Bigger projects are needed—and quickly—to arrest land loss. He's backed by Mark Davis, director of the Coalition to Restore Coastal Louisiana, a conservation group that played a key role in focusing public attention on the land loss problem. "People will wait to act while Rome burns, particularly if Rome is next door," Davis says. "Denial is not an adequate option."

Adding to the sense of urgency is Louisiana's vulnerability to global sea level rise, an issue discussed in detail later in chapters 5 and 6. Scientists predict that the world's oceans will rise by one to three feet over the next century due to the warming of the planet's climate, probably due to greenhouse gas emissions. Southern Louisiana is perhaps the nation's most vulnerable region. (Bahr says the state is America's "poster child for climate change.") Healthy wetlands are the state's best weapon to fight a rising sea.

There is also scientific uncertainty about whether the "big fix" strategy will be as quick and effective at rebuilding wetlands as its proponents believe. The Mississippi built southern Louisiana at an average rate of 500 acres per year, while the artificial diversion and shoreline stabilization schemes promise an average annual return of more than 7,000 acres. Given our record in river system management, some scientists doubt we can out-do nature by more than fourteen times. And while bigger diversions restore land faster than smaller ones, they are not the most cost-effective. An analysis of coastal restoration projects reveals an inverse relationship between cost and effect: on average, a one hundred-fold increase in scale results in a fifteen-fold increase in annual land gain, but at a one thousand-fold increase in cost. You can do more with less, but it takes more time.

And while Bahr and Davis fight for federal funds to address the problem, other state officials continue to make the problem worse.

There's a second element to the land loss crisis. The Corps's levees weakened the bayou by denying it nourishment from the river, but the coup de grace came from the oil and gas industry. After World War II, substantial oil and gas reserves were found under southern Louisiana's swamps and bayous. To access drilling sites and ship the crude they produced, oil and gas companies simply tore deep canals straight through marshes and wetlands. As the number of drilling sites expanded, so too did the network of canals. Today, once vast tracts of wetlands have been diced into little patches by a haphazard network of criss-crossing canals. From the air the canal network resembles a watery spidery web.

Oil and gas companies usually dredged the canals in the cheapest way possible. As they dredged the canals, the mucky soil was simply dumped on the banks of the new waterway, creating tall "spoil banks." These artificial banks acted like tiny levees: They hacked the marsh into tiny patches, each isolated from the surrounding watershed by the mounds of dredged material. Years later the drilling site would run dry and the canal would be abandoned. But the industry didn't fill in the abandoned canals, a simple process that might have given the marshes a chance to recover. Instead, they lobbied sympathetic state regulators to ensure they were never required to do so. On average, 30 to 40 acres of marshes and wetlands are dredged or buried for every mile of canal constructed.

And the canals have a way of widening after they are built. Passing barges create waves that erode the fragile canal banks. The canals themselves allow salt water to intrude farther inland, killing adjacent marsh plants unable to survive in the heightened salinity. Canals typically widen at a rate of 2 to 14 percent per year and in many places have doubled or tripled their width as surrounding marshes turn into open water.

Eugene Turner, a wetlands scientist at Louisiana State University, believes the canals are the primary culprit in wetlands loss. His studies show that wetland loss has been most serious when and where extensive canal dredging has occurred. Local hydrological changes wrought by the canals are the most serious part of the problem, Turner argues, and therefore should be front-and-center in any strategy to combat land loss. Simply backfilling abandoned canals would achieve rapid results and be far cheaper than the "big fix" diversion schemes. "I'm not arguing against diversions," he told me at his office in Baton Rouge. "I'm just saying you need to expand your repertoire to include the canals."

Turner's work created a great deal of controversy in Louisiana. By drawing attention to the oil and gas industry's culpability for land loss, Turner was breaking one of the state's unspoken political rules: The oil-and-gas industry is king and can do whatever it wants. Oliver Houck, director of the Tulane University Environmental Law Clinic, explained it this way: "Oil and gas owns Louisiana. They own the legislature and they own the governor and nothing gets past them in Louisiana politically without their support." So the state has not only refused to make the industry backfill their abandoned canals, they continue to issue more permits to drill more marshes, which makes land loss worse. "It would be the furthest thing from their minds to place the price tag for this problem on the people who caused it," Houck says. So while Louisiana argues for billions in federal dollars to address the crisis, it lacks the political will to make its largest industry do things most other states would have required from the outset.

ॐ

THE DEAD ZONE IS A MUCH MORE complicated problem to solve. There's no controversy about the goal: The quantity of nutrients entering the Mississippi River must be substantially reduced. But there's no "big fix" solution to achieve such results. Instead, the solution will require tens of thousands of tiny actions and investments by thousands of farmers, local and county governments, and individual property owners across thirty-two states. Nobody doubts that it will take decades to address the crisis. What's not certain is whether Gulf ecosystems can wait that long.

There's been a lot of talk about the importance of addressing the Dead Zone, but little concrete action. With thirty-two states and several government agencies implicated in it, the Dead Zone is clearly a national problem. As such it requires strong federal leadership to organize, coordinate, and reconcile the interests, mandates, and activities of dozens of local, state, and national agencies and government departments with jurisdiction over various aspects of the problem. Washington's response has been lackluster.

What little the United States has done thus far was prompted by the Sierra Club (now Earthjustice) Legal Defense Fund, which forced the government to take action under the Clean Water Act. In the end, the government charged an existing organization, the Gulf of Mexico Program, with seeing what progress could be made on the issue. This program has been carefully constructed so as to offend no one. It has no regulatory authority and only a small budget and staff working out of a small suite of offices lost in NASA's sprawling Stennis Space Port complex in southwestern Mississippi. Its membership includes only the five Gulf coast states, even though most nutrient reduction must happen in the Midwest. Industry, the states, some local governments, and a group of federal agencies all participate as program members on a consensual basis. The emphasis is on building trust and cooperation among agencies and agriculture. Progress is slow.

Since its founding in 1988, the Gulf Program has sponsored a mere $12 million in projects, most of which don't involve the Mississippi River watershed. It looks for "win-win" situations in which environmental protection and increased profits coincide. "You're never going to get anywhere unless there's trust," program staffer Hiram Boone explained. Trust is best built on a local level with local problems. "I've met a lot of farmers who have a favorite fishing hole. It may or may not be on their property, but it's special to them. Now for whatever reason they can't get fish there anymore and they want to do something about it so they can take their children or grandchildren there." The program helps out by funding a change in, say, tilling practices or the planting of natural vegetation along the river bank to reduce run-off to that fishing hole.

"It's a grass-roots approach to grass-roots problems," Boone says, local problems fixed by the collective voluntary actions of local people. The Program provides advice and incentive funds, and everybody wins.

To see this grass-roots approach in action I drove up the Gulf coast, across Mobile Bay, and down the Alabama panhandle to Baldwin County, home of the Weeks Bay Watershed Project.

Weeks Bay is the government's model for the future of nutrient management throughout the Mississippi basin. The Weeks Bay watershed doesn't drain into the Mississippi. It's more of a miniature model of the Mississippi. Both drain into the Gulf of Mexico. But while the Mississippi drains half a continent, the tiny streams of this watershed drain but 200 square miles of a single county. All streams eventually empty into two modest rivers, both of which feed into a single estuary, picturesque Weeks Bay, locally treasured for fishing and recreation. Baldwin County is a traditionally agricultural region with rolling fields and woods, sandy shores, and a balmy subtropical climate. It's also the fastest growing county in Alabama as thousands of retirees are drawn to its sunny, quiet, relatively inexpensive, near-rural setting. It leads the state in tourism revenues. As with the Gulf it drains into, water quality in Weeks Bay was deteriorating as increased quantities of nutrients, oil, and sewage drained into its tributary rivers.

Residents of the county wanted to do something to protect their local treasure. Weeks Bay is very human in scale: a mile or so across, 2 miles from north to south, and less than 5 feet deep in most places. It's peaceful and pastoral, with stretches of undisturbed marsh grasses and hardwood swamps interspersed with sandy beaches and modest homes. Locals catch bass and blue crabs in large numbers, and the estuary is an important nursery for shrimp and other fish. People whose families had lived here for generations wanted to keep it that way.

They were greatly assisted in their efforts because Weeks Bay already had the attention of the federal government. As an outstanding example of a healthy subtropical estuary, a portion of Weeks Bay was designated a National Estuarine Research Reserve by Congress in 1986. NOAA maintains a research station here geared towards collecting baseline scientific data for use in coastal management decisions in other watersheds around the country. So when the Gulf of Mexico Program started looking for a site for a showcase pilot project a decade later, they quickly found Weeks Bay.

The result is the ongoing Weeks Bay Watershed Project, the model for future efforts to confront pollution throughout the enormous Mississippi watershed. It's a project largely by and for Baldwin County residents, but the county has substantial financial and research assistance from the Gulf of Mexico Program and eighteen other local, state, and federal organizations. A comprehensive, basin-wide management plan has been drawn up and dozens of small-scale, grass-roots projects are underway to begin the slow, piecemeal task of reducing nutrient loading and other pollution running into the bay. This model for stopping the Dead Zone stands in stark contrast to the top-down, "big-fix" strategy to save Louisiana's bayou country.

Eve Brantley took an afternoon to show me around the rural back roads of Baldwin County. Eve was the Research Reserve's gregarious young outreach coordinator. She must be good at her job: Although a recent transplant from Georgia, Eve already seemed to know half the people in the county. We drove through pecan orchards and cornfields, down red dirt roads and quiet country lanes, Eve waving to passers-by as we stopped here and there to see pilot projects in action.

There was a cattle ranch with a stream running through it. The rancher's cows had liked to fight the summer heat by standing in the stream for much of the day. There they had defecated in bovine quantities, making the stream one of the worst polluted in the watershed. Now the cows looked longingly at the stream from behind fences the rancher had erected with Weeks Bay Watershed Program funding. A few miles away a soybean farmer had changed her tilling practices to reduce rainwater run-off into a nearby river. The Program had helped a local family replace their leaking sewage tank with a state-of-the-art system that incorporates a small artificial wetland to cleanse effluent before it can leak into Weeks Bay a few hundred yards away.

"We tell people that anything that ends up in the water winds up in the Bay," Eve says. "People care about that. Even more so if you can offer some cost-sharing funds."

The Gulf of Mexico Program's strategy is to set up hundreds of local watershed programs throughout the Mississippi watershed. Residents of small, county-sized tributary watersheds could coalesce around protecting their local river. Farmers might engage in precision agriculture or alternative tilling and drainage practices, or agree to reestablish wetlands or natural riverbank vegetation. Public education campaigns might reduce illegal dumping or excessive use of lawn fertilizers and weed killers by suburban homeowners. Matching funds might provide incentives for municipal governments to improve wastewater treatment or for chicken farmers to better treat or dispose of manure.

But if such an approach is to work, it will require a far more active commitment by federal and state governments. Most watersheds are not as gosh darn lovable as Weeks Bay. Nor do they happen to have a National Estuarine Reserve in place to protect habitat and serve as a focal point for watershed activities. Many counties lack Baldwin County's financial resources. Nor will it be so easy to find public consensus in, say, Mississippi's politically backward delta country or among urban residents in St. Louis or Denver.

Without strong leadership and financial support, this strategy is doomed to failure. Water quality will be improved in certain places, which is wonderful. But at the current pace, nutrient reductions would be lucky to keep pace with the growth of the American population and economy. A significant reduction in nutrients reaching the Gulf—say 40 or 50 percent—simply requires a much more active effort. Without such a reduction, the Dead Zone may never go away.

Even under the most optimistic forecasts, the Dead Zone and wetlands loss won't be undone for many decades. Nobody knows how much permanent damage will have been done to life in the Gulf of Mexico by then. The fear is that these and other stresses may push the ecosystem over the edge and into collapse. Then America might have its own Black Sea.

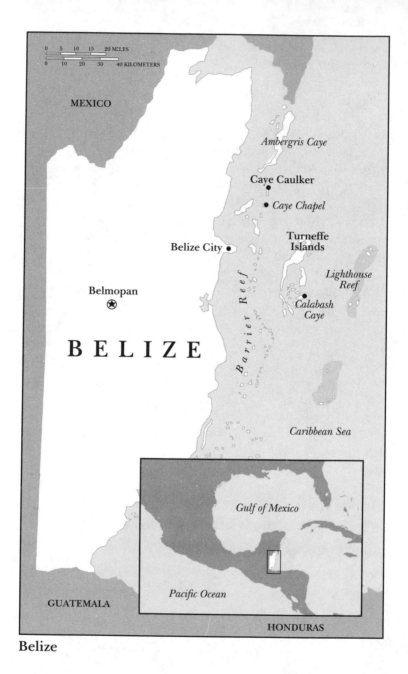

Belize

five
Fall of the Magic Kingdom

 CAYE CAULKER IS AS RELAXED a tourist destination as you can find. It's a low sandy island, 5 miles long and one-quarter-mile wide, except at the sparsely populated southern end, where thick mangroves triple its girth. The village clusters around two parallel tracks of soft coral sand—the "front" and "back" streets—named for their proximity to the trade winds. There are no cars, just a couple of rusting pick-up trucks to move cargo and a small fleet of electric golf carts. Most people simply walk. Nothing is very far away on Caulker, and nobody is in any hurry to get where they're going.

The handful of restaurants, guest houses, and family-owned hotels cluster on the front-side shore, facing the open sea alongside the homes of the more influential islanders and a handful of U.S. and North European–born Caulkerites. Most of the latter came for a visit a decade or more ago and never left. All front-side property is cooled by the steady winds, which blow hungry mosquitoes and sand flies to the back side of the island, where they bite less fortunate Caulkerites sitting in still, fetid heat a mere hundred yards away. The back shore is still and sweltering day and night, its homes facing towards the thick jungles of mainland Belize, the urban jungles of Belize City, 20 miles across the waves.

Like most visitors, I had come to see Belize's world-renowned coral reefs. Belize itself is something of a backwater: Few Americans have heard of this little nation of 180,000 people; even fewer could find it on a map, a New Hampshire–sized plot of land nestled between Guatemala and Mexico at the base of the Yucatan Peninsula, facing the shimmering sea. But as other parts of the Caribbean are marred by uncontrolled development, more and more U.S. and European travelers are discovering Belize, a former British colony that hasn't lost its laid-back, rough-around-the-edges, "what's-the-hurry, man" character. And a great many of these visitors are underwater divers, drawn to Belize's barrier reef, the longest and perhaps most beautiful in the Western Hemisphere. These reefs, which extend in a nearly continuous offshore ridge along the entire length of Belize's 180-mile coast, are among the last in the Caribbean basin that remain in more or less pristine condition. As more and more foreigners flock to the reef, island communities like Caulker are experiencing fundamental, irreversible change.

Until recently, Caulker was a fishing community. Its nine hundred residents fished the reef for sustenance and earned their cash income trapping the spiny lobster, a large, clawless cousin of the American lobster said to have been so common it could be herded onto Caulker beaches with palm fronds. Today the lobster population is dangerously low, in part because many island restaurants buy illegal, undersized lobsters, encouraging their capture by poor fishermen. Juvenile lobsters, groupers, snapper, and other reef creatures are served in diced form—lobster jambalaya is especially common on Caulker menus these days—to conceal their illegal size. My host told me that when he first came to Caulker two decades ago you'd be served large fillets cut from large reef fish. "Today you're served a whole fish, and you're getting less of a meal. There just aren't any big fish left to catch."

So Caulker's survival increasingly is tied to tourism, catering to backpackers and divers fleeing resort developments on crowded Ambergris Caye 12 miles to the north. Former fishermen work as boat drivers, tour guides, divers or dive masters, as bartenders and carpenters. Those with money—the island's traditional leading families—invested in restaurants and small hotels. Those with neither money nor connections are forced to continue fishing, with ever diminishing returns, though a few fish by choice, preferring to work on the reef. A generation ago Caulkerites' lives were entwined with those of the reef inhabitants; now they're tied to trends in international tourism, to foreign stock markets and the price of aviation fuel. Yet tourism, like fishing, depends on the reef. That much hasn't changed.

But the reef has.

Corals, the tiny anemone-like animals that built the reef and, by extension, Caye Caulker itself, can only thrive in water that is clean, clear, shallow, and warm. Even in their natural state, Belize's mainland rivers and streams accumulated sediments and organic material as they flowed through the dense jungles of the interior. But along a now submerged ridge several miles offshore corals found conditions perfect for their growth. Layer upon layer, generation upon generation, the tiny coral polyps excreted limestone shells around their bodies, slowly building the barrier reef. For many thousands of years the corals thrived, enduring rising seas and changing weather while providing fish and raw materials for ancient Mayans and Spanish conquistadors, English pirates, and modern Belizeans.

But tourism has triggered a development boom along the barrier reef, especially in the area around Caye Caulker. The next island up the reef from Caulker, Ambergris Caye, is the epicenter of the nation's tourist industry; over a third of all foreign visitors to Belize stay on Ambergris during their visit, and many go nowhere else. Ambergris's main town, San Pedro, is now an affluent resort catering to North American package tours. Its forests are being cleared to make way for more hotels and retirement homes, and a 200-acre resort development with an airport, marina, equestrian center, and gambling hall. Sewage, pollution, and soil erosion are on the increase, complicating the lives of coral polyps and other creatures on the reef, which were already disturbed by crowds of divers and snorkelers.

The island to the south of Caulker, Caye Chapel, was once among the most beautiful in Belize, a 2-mile-long, cigar-shaped island with wide sandy beaches and tranquil groves of palm trees and mangroves. Now it's a depressing sight. The owner, Kentucky coal mine operator Larry Addington, converted the entire caye into a sprawling eighteen-hole golf course and resort complex. Never mind that the cayes are desperately short of water, that the surrounding reefs can't tolerate run-off laden with pesticide and fertilizer residues, that the mangrove forests they cleared were the main thing holding the island together. Nor that the mangroves are so important to fisheries that it is now illegal to clear them at all. The developers built their eighteen holes. Then they wrapped half of the island in a girdle of landfill and steel seawalls to keep it from subsiding away into the sea. They built desalination plants to produce water to irrigate fairways and greens as they baked under the tropical sun. Locals told me they thought the brine from the desalination equipment was harming nearby reefs. Development has slowed only since scientists at the Belize Audubon Society, an organization that manages many of the country's national parks, convinced the government to review the environmental impact of plans to expand Chapel's airfield by dredging sand from the surrounding seafloor.

Then there's Caulker itself, which appears to be going the way of Ambergris. Every year new hotels pop up along the windward shore, where mangroves are replaced with ugly retaining walls. There's a gigantic, half-completed concrete resort on the north end of town that looks like it's about to follow the collapsing shoreline into the sea. To the south, mangroves have been replaced by a little-used airstrip and another abandoned resort complex. The island seems cursed with developers savvy to neither environmental nor business realities. Nonetheless, the island has more inhabitants and more visitors every year, and they leave an ever-larger environmental footprint.

Beautiful reefs attracted tourists and developers to these islands. Some island residents have realized that they are squandering the very thing that, one way or another, puts food on everyone's table: the living reef. But their early efforts to protect the corals have encountered some unexpected difficulties.

❧

JAMES AND DOROTHY BEVERIDGE came to Caulker in 1980, abandoning northern California and a consumerist lifestyle that no longer made sense to them. Jim set up a successful underwater photography enterprise, and together they operated a small photography gallery and book shop for the tourists. Recently they closed the shop, and word around town was that the lease hadn't been renewed because somebody on the island didn't want the competition. I'd arranged to meet Dorothy at the local office of the Belize Tourist Industry Association (BTIA), where she now works organizing the hiring and support of a warden for the newly established Caye Caulker Marine Reserve. We met long after sunset and it turned out that the office was set among the mangroves a mile out of town on a tiny front-shore footpath. The path was hard to follow in the pitch-black of night. I kept wandering off into the bushes, and even when I stayed on the track I startled groups of mangrove crabs or solitary night herons on the trail. But inside the BTIA office it was bright and cozy, with a half dozen busy people using e-mail or the Internet on one of several computers. Dorothy, a healthy middle-aged American in T-shirt and khakis, greeted me at the door and we sat down at one of the long worktables to talk about the reef.

"The year we arrived was the first year that Caulker had twenty-four-hour electricity, and that was mostly for the fishing community and their freezers," Dorothy recalled. "Then they started bringing in TVs and VCRs to watch tapes from the U.S. In the late eighties they started getting telephones. Then the ladies got washing machines so they didn't have to use washboards anymore—but that happened *after* televisions and VCRs. Development was a step-by-step progression because once you had electricity you could build larger hotels and resorts. Around 1992–93 things pretty much switched over to tourism. People just saw that there was more money—and it was a lot easier to build and operate a hotel than to go fishing every day."

The Beveridges also noticed changes in the reefs as Jim photographed them. Growth and development were causing them to deteriorate. Developers determined to bring golf to the cayes had dredged the north and south ends of Caye Chapel, and the sediments smothered the elkhorn coral over the entire area. Every year more of Caye Caulker's mangroves are cleared, triggering serious erosion that threatens to smother nearby reefs while eroding newly cleared real estate. Then people build sea walls where the mangroves were, which often simply displaced erosive waves and currents farther down the shore. "There's a national law that says you need a permit before you can cut any mangroves, but in practice out here people just cut them down if they're in the way. There's only one mangrove person in the ministry for the entire country and he only got a boat recently. Belize has a lot of good laws on paper, but it's a matter of educating people so that they understand why these things should be protected.

"If you cut down all the mangroves there'll still be fish on the table because of Hick's [Caye] and other cayes. But if you don't have mangroves, you won't have an island. A lot of people think you can just build sea walls and that will solve the problem. But the mangroves are part of the reef, and the reef protects and maintains the island.

"Fortunately there've been good things too. I think Belize has done a really good job of building environmental consciousness on a grass-roots level. People now understand that their livelihoods depend on having healthy reefs, and 'environmental education' has become a catch-word. There's a growing understanding that environmental protection isn't separate from cultural survival and sustainable development. But there's still a lot to be done."

∞

FEW PEOPLE HAVE DONE MORE for Caulker's reef or raised more controversy doing it than Ellen McRae, an American-born marine biologist who has lived in Belize since 1982. She's married to Orlando Carrasco, one of the founders of the Fishermen's Cooperative and a member of one of the old Caulker families, and they have a young son. Ellen can't be brushed aside as easily as most outsiders. She's family on Caulker, and she's willing to fight for what she believes in: that the reef should be protected for the long-term good of the island and its people. When her name came up in conversation, many people would stiffen or roll their eyes slightly. I gathered that she'd stepped on more than a few toes in her long battle to establish a marine reserve on Caulker. Idealists aren't always the most diplomatic negotiators.

But the woman who greeted me at the door of the weathered wooden house on a back street was a warm, stout, unpretentious person in an oversized T-shirt, ankle-length dress, and bare feet. "Just have a seat on the steps for a minute," she said. "I've got to get my laundry off the line."

A few minutes later we were seated at her kitchen table, which was concealed under an enormous pile of maps, documents, magazines, and newspaper clippings dealing with every imaginable aspect of marine conservation. Here at this table, Ellen had waged war with developers, politicians, government bureaucrats, and other islanders in an effort to establish what is now called the Caye Caulker Marine Reserve, recently passed into law by the parliament in Belmopan. On paper, at least, 100 acres of mangrove forest at the north tip of the island and several square miles of barrier reef are now under protection, although what sort is not clear as yet.

"It's a Pyrrhic victory. You've never really won because there are always vultures around," she said wearily, pausing occasionally to breathe from an anti-asthma inhaler. "I don't want to insult your politics, but nowadays people watch TV and tapes here and they see all this Reaganism, this rampant consumerism. Hey, they all want that. They want to be 'Big Time' now. In the old days everybody was helping each other out on Caulker. Look at this house. Hurricane Hattie moved it two hundred meters [in 1961] and people and friends came and helped move it back into position. That wouldn't happen today. No, now it's all about money, and people who have it want more. Money is like a drug, and if you stand between people and their drug you can get into some serious trouble."

McRae soon discovered that the marine reserve was getting in the way of some powerful people. As first planned, the reserve was to include a large part of the thick mangrove forest on the then-undeveloped southern end of the island. But, as McRae says, "money talks," and pretty soon plans for the reserve were put on hold. Next thing she knew, they were building an airstrip on the southern part of the island, triggering a new property boom there. "After that, they're hiring people to fill in the mangrove swamps around the landing field so they can build a resort hotel or something," she said. "I mean it's filled with endangered crocodiles and I happen to like them. I held a baby crocodile when we had a scientific expert here a few years ago—nice little creature, I really like them."

She took a hit from her inhaler. "Sorry. Anyway, back in 1993 they were already starting to fill the mangroves because of this project run by an American developer from Kansas. It's terrible. When you wipe out the mangroves, the beach erodes. You lose the nursery ground for lobsters, snappers, and other fish. The sediment starts following onto the reef. One day I'm out there in the mangroves with Mr. Kansas, trying to show him some of this [environmental] stuff. We're out in the middle of nowhere in the forest and he says to me very pointedly 'you know, it would take a long time to find a body out here in these mangroves.'"

In the end the airport wound up where the mangrove reserve would have been. Few planes land on the airstrip since tourists seem to prefer to take the boats from Belize City. The resort development has failed for now, remaining nothing more than a cluster of half-completed concrete shells at the side of the runway. The reserve was reduced in size and scale and relocated to the narrow north end of Caye Caulker. McRae says that despite the failure of the airport resort, five of the island's leading families have plans to build five more resort villas on the north end, threatening to damage the marine portion of the reserve through pollution and soil erosion.

"Its hard to fight for democracy," McRae says with resignation. "I've given so many hours of my life for the past nine years. If I'd known how extensive this would have been I probably wouldn't have even started it. I dropped my own work to try to make this reserve happen." She shook her head. "So we've got a marine reserve," she said. "Now the question is what we plan to do—or not do—with it."

∞

ON PAPER, BELIZE HAS TAKEN IMPRESSIVE STEPS to protect the health of its reef system, which has been named as one of the "seven underwater wonders of the world" and is currently under consideration to become a UNESCO World Heritage Site.

Part of the reason this poor developing nation has taken action is because its leaders have recognized the importance of the reef to its economy and national development plans. In the last two decades of the twentieth century, Belize's economy ceased to rely on the sale of citrus crops to foreign nations as growing numbers of foreign tourists came to the country to see its largely undisturbed jungles and, most of all, the barrier reef itself. The number of tourist visitors doubled from 1988 to 1995 and now accounts for 15 percent of its Gross Domestic Product (GDP) and more than a quarter of export earnings. (Fishing—also dependent on healthy reefs—accounted for another 3 percent of GDP and over 5 percent of exports.) By 1997, almost as many tourists visited annually as there are Belizeans, and most spent the majority of their time on the coast. Almost a third came to dive and two-thirds snorkeled on the reef during their stay.

"The challenge is for Belize to find a balance between conservation and development," Valerie Woods, deputy director of the Belize Tourist Board, told me in her Belize City office. "Nobody's going to come to an area that's overused, abused, and saturated with visitors. Today's travelers are more appreciative of the need to protect the environment. Pristine reefs and rain forests are harder and harder to come by, so ours are only going to become more valuable as time passes."

Since gaining independence from Britain in 1981, Belize has established an impressive system of national parks, forest reserves, wildlife sanctuaries, archeological sites, and nature and marine reserves. Today 30 percent of Belize's land area is under some sort of legal protection. Nine reserves protect various parts of the reef system, including the new Caye Caulker Marine Reserve. Another eleven marine protected areas are under serious consideration, including Calabash and the Turneffe archipelago. Development decisions that affect the coastal zone are to be coordinated under a single integrated Coastal Zone Management Agency. Parliament has passed laws protecting mangroves, marine life, and reef systems.

But Belize is a poor country with limited human and financial resources. There's income inequality, chronic unemployment, and widespread poverty. Lacking the funds and personnel to manage the reserves, the government has formed partnerships with conservation organizations, universities, environmental groups, and development agencies in Belize and developed countries. The newest reserve, Caye Caulker, is to be managed and administered by the community itself. But the majority of reserves and protected areas remain without administrators, wardens, or management plans, some without any oversight or administration at all. They exist on paper, but in practice, nobody is watching.

"A lot of work needs to be done to make most marine protected areas functional," said Valdemar Andrade, advocacy coordinator of the Belize Audubon Society, a nonprofit organization that administers twelve parks and protected areas on behalf of the government. "If it includes a caye, the reserve falls under the jurisdiction of one government ministry, while the underwater portions fall under another. The marine portions are divided up into use zones—fishing zones, scientific zones, protected areas, recreational zones—but you need to draw up a management plan to coordinate these things, and wardens to enforce them or collect use-fees. Many of the country's protected areas don't have these things."

Even those organizations with some management resources face regular difficulties. The Belize Audubon Society manages the country's two best-known, most-visited protected areas: the Hol Chan Marine Reserve off Ambergris Caye and the Half Moon Caye National Monument, an idyllic coral caye near the Blue Hole on Lighthouse Reef.

Half Moon's operations almost break even since the Blue Hole attracts so many divers, who visit Half Moon during compulsory dive intervals. But the monument only protects a tiny fraction of Lighthouse Reef, and the two wardens who live on the caye aren't able to protect it from many stresses. The island is the breeding site for the rare red-footed boobies, which live up to their names: Nesting boobies regarded me with bovine nonchalance even when I was standing only a few feet away. The recent introduction of rats by visiting ships threatens to undermine these placid birds. I found nearby protected reefs to be among the most fabulous diving of my visit, but nearby pollution, anchor damage, and overfishing affect the protected parcels.

"We hope to eventually create a body that will manage the whole reef system," said Belize Audubon Society executive director Osmany Salas. "All the interest groups support this—fishermen, tour guides, and the Park Authority. The problem is money. Foreign lenders would have to become involved because there isn't this kind of money in Belize."

The problems are worse in the Hol Chan Marine Reserve, despite the presence of a management plan and hordes of fee-paying visitors from Ambergris and Caulker. The reef wall looks healthy enough, but when I went diving amid the crowds of boats in the shallows of the reef cut the impact of tourism and nearby development was clear. No sooner had I rolled into the water than I was surrounded by huge numbers of fish and nurse sharks waiting for handouts of food. Several large Crevalle jacks became very agitated when I failed to deliver a bit of lunch, repeatedly swimming in front of my mask to get my attention. I watched an impatient nurse shark poke its head out of the water next to an adjacent boat, trying to attract the notice of the pasty-white snorkelers eating lunch on the stern. Tour guides feed the fish to give their clients a good show. This practice skews the ecosystem by supporting more aggressive species over their competitors and prey. It also endangers snorkelers and divers themselves by training sharks and predatory fish to regard people as a source of food. The glint of a ring, belt buckle, or dive watch can stimulate an unprovoked attack.

The reef crest and wall around the cut made for a depressing sight. The vast majority of corals had died and the wall was now covered in an expansive rug of kelp and seaweeds. Divers and snorkelers can cause considerable damage merely by touching reefs, which removes their protective mucus, making them vulnerable to disease, predators, and bleaching. Unfortunately, ignorant visitors often sit on reefs, walk on the coral heads, and even break off pieces of coral to keep as souvenirs. But these are merely contributing stresses. Sediment and nutrient run-off from Ambergris, Caulker, and even far away Chetumal, Mexico, conspire with mangrove destruction to erode these shallower parts of the reserve. It's probably piling straws on the outer reef's back, pushing it closer and closer to the crash point.

∞

ASK AROUND ABOUT BELIZE'S CORAL REEFS and one name comes up over and over: Janet Gibson. Born in Belize and educated in the United States, Gibson has worked with successive Belizean governments to increase protection for reef systems and to improve conservation of the country's bountiful living natural resources. She was at the forefront of the effort to make Hol Chan a marine reserve. For the past five years she has headed the Coastal Zone Management Unit, a special administrative office in Belize City financed by the Global Environment Facility of the United Nations Development Program (UNDP).

"For we Belizeans the reef is important for all sorts of reasons," Gibson explained, sitting at a desk surrounded by satellite images, topographic maps, and piles of reports. "The majority of tourists come here because of the reefs. Most of the species the fisheries rely on live on the reef. The reefs protect the coast from storms—something that will become more and more important if sea-level rise occurs. It's a part of Belize. It's important for the whole world as well because it's the longest reef system in this hemisphere. The pharmaceutical industry is actively looking for medicines in sponges, soft corals, and the living reef corals. Then there's the sheer beauty—the relaxation that comes from visiting this undersea realm.

"So people ask what we can do to protect it. Well, we went through a series of evolutionary steps in our efforts to conserve the reef. At first people tried to protect one species at a time, but quickly realized that to do this you must protect its habitat and ecology. So then we established marine protected areas like Hol Chan. That does good things. If you place them correctly you can protect a range of habitat types, conserve nursery areas and spawning areas of reef fish, or protect an endangered species. But they also work as fish sanctuaries. We found that if you close one area to fishing, catches increase in adjacent areas. But now we realize this is not enough.

"To protect the reefs—or any other ecosystem for that matter—you have to take a broader approach: an ecosystem approach. You have to step back and look at all the impacts—those from sewage or wastewater, from industrial pollution and the run-off from agriculture. This last part is particularly important here in Belize since our citrus and banana farms use enormous amounts of fertilizers—especially in the new farms that are now spreading to less fertile areas that need more fertilizer to make the crops grow. You need to look at pollution in the watersheds, at coastal construction and urban expansion, at fisheries and forest loss, at the effects of tourism and air pollution. You have to take the 'Coastal Zone' approach, which is why we worked to create this office.

"Without the Coastal Zone Management approach we'd started to see impacts that were beyond our control. A reserve can be damaged by sedimentation from a nearby development, by the collapse of an overfished species outside the reserve, or by algae blooms triggered by nutrients flowing down a river from a fruit plantation. If there isn't any coordination between economic sectors your efforts to conserve an ecological system are not going to be successful."

So Gibson's office works to coordinate policy between water management authorities, fisheries representatives, and the forestry ministry. They sponsor public education programs to show people how the mangroves, reefs, fisheries, and cayes are all part of the same system, how each must be conserved if people who rely on them are to thrive.

The project would never have gotten off the ground without several millions of dollars of support from UNDP and other foreign sources. But now that the office is up and rolling, successive governments have given their firm support to its efforts. When I visited, Belize had just undergone a change of governments for the third election in a row, the traditional opposition having trumped the customary ruling party for the second time. They agreed on very little, except the need to protect the reef. Perhaps this is one thing Belizeans can rally around regardless of class, race, or party affiliation: Everyone's future is tied to the health of the reef, the goose that lays Belize's limited supply of golden eggs.

〜

I SPENT SEVERAL DAYS DIVING ON THE BARRIER REEF around Caulker and Hol Chan and it was easy to find signs of deterioration, especially in the shallower, more protected areas inside the reef crest and around the heavily visited Cut. In deeper water the reefs looked better, although there weren't as many large fish as I had expected. As I'd never explored Caribbean reefs before, I began wondering what the reef was supposed to look like.

For comparison, I thought it would be a good idea to visit some of Belize's allegedly pristine reefs. Not surprisingly the best ones turn out to be those farthest from the coast, where they're concentrated in three ring-shaped island clusters well outside the barrier reef, 20 to 40 miles out to sea. One of these is the Turneffe archipelago, an offshore chain of mangrove-covered cayes surrounded by its own barrier reef system. The University College of Belize monitors the health of Turneffe's corals from a modest research station on Calabash Caye, a small island on the far side of the archipelago. The station's director, Vincent Palacio, had kindly invited me to come out and have a look.

Arranging to get to Calabash from Caye Caulker hadn't been easy. Few travel to this part of Turneffe—there are no settlements, few people, and it's far from the routes to Belize's premier offshore dive sites. In the end I succeeded in chartering an outboard skiff piloted by Mark, a young Caulkerite who made a living running water taxis to and from Belize City, 20 miles west of Caulker.

I was in Belize with Shep Smith, a master diver and childhood friend who commands a government survey vessel on Chesapeake Bay. Shep had enthusiastically agreed to spend his vacation making sure I didn't drown amid the breathtaking splendor of Belize's coral reefs. On this particular morning, we found ourselves confronted with a new drowning risk. Shep looked over Mark's open skiff, smaller than the lifeboats on his command, smiling and shaking his head. With only two passengers, chartering a larger boat on the island proved out of the question. But the weather was agreeable that morning and Mark told us not to worry.

He was right; other than a glancing, full-speed collision with a mangrove tree while negotiating a narrow passage on Turneffe, we had no trouble getting to Calabash. We raced south for 30 miles down the length of the barrier reef to minimize the time spent in "the blue," the open Caribbean on the reef's other side. First Caulker and its tiny tourist hamlet vanished astern. Then we passed Caye Chapel, denuded and depressing, its absurd golf course baking under the tropical sun. But south of Chapel the cayes remained in a more natural state, most completely covered in dense green mangrove swamps.

Mangrove trees have the remarkable ability to grow in salt water and do so on tropical lowland coasts worldwide. They fill the same ecological niche as do saltmarshes at higher latitudes, building and stabilizing the landscape, cleansing run-off from the land, and providing key nursery habitat for all manner of marine creatures. They grow in dense colonies, their spidery roots actually reaching out of the water and into the air to capture carbon dioxide for the roots below. An island of mangroves looks like a hardwood forest that has been flooded up to its leafy crowns. It's impossible to tell where the sea ends and dry land begins—or even if the land begins at all.

That's what the islands of Turneffe are like: swampy green conglomerations of vegetation that may or may not cover real soil. Those that do, like Calabash, owe their terrestrial state to centuries of work by the mangroves, endlessly capturing sediments in their rooty webs. The archipelago is constantly shifting as the mangroves shift and grow around the massive Central Lagoon. A placid tropical lake two miles across, the lagoon's shores are devoid of farms, villages, any human structure at all. Even at the surface the lagoon teems with life. Herons and ibis wade in the shallows. Cormorants nest among the trees between offshore fishing expeditions. Frigate birds—a magnificent soaring species that steals food in mid-air from smaller birds—nest so thickly on the mangroves that some island canopies are completely smothered with the creatures and their droppings.

Calabash lay just outside the ring of lagoon islands, its perfect sand beach facing the trade winds whipping across the warm Caribbean. It was the perfect place to study corals. Belize is one of the few countries in the Caribbean whose reefs are still regarded as "semi-pristine," thus far spared the worst excesses of developers, fishermen, and human reproduction. Such stresses are increasing inside the Barrier Reef as Belize City and the tourist enclaves of Ambergris and Caulker expand. But Calabash Caye is far from the sewage outfalls and golf courses, the marinas and airports. Nestled in thriving mangroves and open to the cleansing circulation of the wider sea, Calabash offered a chance to see healthy, unspoiled reefs—a basis of comparison with the dying corals near Ambergris and elsewhere in the Caribbean.

Or so I'd hoped.

☙

DR. PALACIO, A LARGE AND GRACIOUS Belizean man wearing a Chicago Bulls T-shirt, greeted us warmly as we tied up at the Marine Research Center's modest dock. He placed us in the capable hands of Jonathan Kelsey, an American marine biologist and Peace Corps volunteer who was helping manage the Center's coral assessment program. The Center—a small compound of wooden buildings next to the beach—housed scientists and volunteer divers who for months had been surveying the health of the local reef as part of an international data collection effort. Together with more than two dozen other marine laboratories in the Caribbean, the Center forwarded its standardized reef assessments to Jamaica's University of the West Indies for detailed analysis.

Unfortunately, Jonathan found himself in labs and offices more than in the water and was pleased to have a professional excuse to get into his SCUBA gear. The plan was first to snorkel around the shallows inside the reef, then take a boat farther out and dive the outer reef wall. We expected both areas to be in excellent condition.

Mark's boat was so small that we elected to use the Center's dive boat, a larger skiff now filled with diving gear. Joining us were Shelly Jeps, a Canadian biologist interning at the Center, and the Center's coxswain, Jimmy Sanchez, renowned for helping scientists wrestle live alligators into this very boat. As we pulled away from the dock, Mark and Jimmy quickly became engaged in friendly but incomprehensible conversation. Many Belizeans are descendants of Africans brought to Barbados and Jamaica as slave labor on British and Spanish sugar plantations and speak a Caribbean form of pidgin English I never got a proper grip on. I'd overhear a phrase like "Falla-foot jumbie, you jump eena water" (If you follow a ghost, you'll end up in the water) and spend hours afterwards puzzling over its meaning. Back on Caulker I'd asked a dive master what the dialect was called. "It's called *good* English," he said with a confident smile. "What I'm speaking for your benefit is *bad* English."

A quarter-mile off the beach, Jimmy idled the engine and we drifted and Jonathan, Shelly, Shep, and I strapped on our fins and rolled into the shallow turquoise water. We were a thousand yards inside the protective wall of Turneffe's outer reef. Its crest was a few feet underwater, but I could see the white surf crashing over it in the middle distance. Just a little before that line of surf, the reef rises from the shallows like a living fortification, protecting the island from the fury of the open Caribbean.

The sea was so warm none of us wore a wet suit. Unencumbered by SCUBA gear, we could move quickly over the shallows, and I ducked my face under the surface to survey the corals below.

It was an alarming sight.

I was floating over a gently sloping plane of sand covered in coral heads and stretching as far as the eye can see in the crystal-clear water. It was clearly a rich coral environment. They covered much of the bottom in rocky formations of all shapes and sizes: egg-shaped boulders the size of lounge chairs, coral bouquets that look like moose antlers, great thickets of antler-like finger coral, and a huge ornamental "tree" of elkhorn coral, its branches casting shadows in the sand beneath its base. Colorful sponges, sea fans, and schools of tropical fish completed what should have been a most beautiful scene.

Only something was very wrong. The coral heads should have been rich in color: brown elkhorn coral, golden and mustard-yellow star corals, corals in greens, blues, browns, and reds. Instead, bright white boulders dotted the seascape in all directions, a sign of severe coral distress. A centuries-old stand of elkhorn coral as big as an elephant was now dead and smothered in a thick two-year growth of brown algae. We found long stands of finger corals, normally yellow and brown, now bleached bone white. Across the plane, the corals appeared to be dying.

Jonathan was shocked. Just two weeks before he had snorkeled in this same spot and all the corals had been healthy. In past years the University had searched in vain for examples of bleached corals to show interested guests. Now most of the back-reef plane was dotted with white, which for corals is the color of death.

We climbed into the skiff and Jimmy swung us onto a northerly heading and opened the throttle all the way. The boat pounded through growing wave crests as the rest of us prepared our dive gear.

Cradling a yellow oxygen tank—noticeably newer than the ones I'd been renting that week in Caulker—I locked down the nylon straps attaching it to an inflatable vest called a "B.C.," short for buoyancy compensator. I attached the fitting to the tank valve that distributed compressed air to my breathing regulator, depth and pressure gauges, back-up regulator, and B.C. Holding the gauges away from my companions—they sometimes explode when first pressurized—I gingerly opened the tank valve, filling the hoses with a sudden hiss. I glanced at the gauge: 3,000 pounds per square inch. It's the right pressure for once; every tank I'd rented in Belize up until then had been filled well beyond its rated capacity.

I took a breath from each regulator. Cool, dry air filled my lungs. Inhale, exhale through the regulator: I sounded like Darth Vader. I pressed a button to inflate the B.C., which expanded obligingly like a balloon until the safety valve opened with a reassuring "pop" to prevent the vest from overfilling. I pressed another button to empty the B.C. Under water, pressure increases with depth at a rate of one atmosphere every 33 feet, reducing the body's buoyancy. A diver can compensate for this effect by inflating or deflating the B.C. with depth, and, with practice, can maintain neutral buoyancy. Properly adjusted, you float effortlessly at a chosen depth, slowly rising and sinking as you breathe in and out. Approaching a small boulder or sea fan you can fill your lungs with air, rise gently over the obstacle, then slowly sink again as you empty buoyant air from your lungs. It's a little like becoming a hot air balloon.

We would be diving on the outer reef wall near Blackbird Caye, Calabash's immediate neighbor in Turneffe's chain of islands. A mile off Blackbird coral beach, Jimmy cut the engine and the four of us put on masks and fins—these already dry although we'd finished snorkeling only fifteen minutes before. Sea surface conditions were flat so we'd be making a one-way trip down the reef while Mark and Jimmy followed our surfacing bubbles with the boat. To orient myself, I asked Jonathan what the reef topography would look like. "Don't know," he said. "I don't think anyone's ever been diving here." With that he sat on the gunwale, held his mask to his face with one hand, and rolled backwards into the turquoise water.

A few minutes later we were all bobbing in the water a few yards from the boat, B.C.s inflated for flotation. Watches and dive computers were set and, with a thumb's down signal from Jonathan, we emptied our vests and began sinking slowly down into the realm below.

With a soft kick, I rolled slowly into a horizontal position, facing downwards like some slow-motion sky diver. I was gently falling over a mountainous landscape, a living coral range running north-to-south in parallel with the caye shores hidden somewhere half a mile to my right, Turneffe's protective barrier reef. The sloping wall was some 50 feet below, steadily advancing, its details emerging from the soothing blue twilight. Buttressing coral ridges reached out into the deep like claws, forming dramatic canyons between them. Beyond these ridges the reef wall dropped off in sheer cliffs like the edges of a bottomless pit. Recreational divers venture no farther than 150 feet beneath the surface; the drop off here fell to more than 1,000.

I rolled head-over-heels and looked up at my bubbles rushing towards the surface 20 feet above. The sun shone through a kaleidoscope sky, the shimmering plane of surface waves seen from underneath. In clear water, it is a beautiful sight, this delicate barrier between air and ocean. Life began on the liquid side of the barrier. It took billions of years of evolution for our ancestors to break through to the other side, seizing primordial mud and tide pools, swamps and riverbanks, later the forests, plains, and mountaintops. Only in the past century have we hominids started coming back, visiting the mother-realm with the help of aqualungs, bottled air, masks, fins, and dive computers. We've changed a lot in the past 360 million years and we feel more than a little out of place here. But most of us who make the return trip keep coming back for the rest of our lives.

I did another slow-motion somersault to confront the advancing landscape. I was falling onto the gently sloping face of the reef crest, a wall created by untold generations of tiny coral polyps. The current generation covered the landscape in an array of contour and structure: yellow and blue boulders; great brown "moose antlers" of elkhorn; stubby pillars in reds and oranges; occasional boulders shaped like disembodied brains as large as a diningroom table; purple sea fans, their enormous "leaves" an intricate lacework that sieves passing plankton from the gentle underwater wind. A few feet from the bottom I began letting air into my B.C. in short bursts, checking my descent. I hung motionless over a purple barrel sponge so large I could have climbed into it. *Houston*, I thought, *the Eagle has landed.*

A flock of a dozen or more curious yellowfin snappers were checking me out, swimming with me in slow circles, peering through my mask with big unblinking eyes. I gently kicked my feet, careful not to touch the barrel sponge below, and drifted over a bouquet of purple coralline fingers. An inch-long damselfish shot out from an unseen crevice, took an aggressive lunge towards me, then shot back beneath the corals like an upset dog. A moment later it charged me again, then took station staring up at me, mouth opening and closing with agitation, and lunged again as I continued to drift past. Just another enormous intruder to keep away from its grazing territory. Despite our bubbles, our unusual appearance and scent, most reef fish treated us as if we were just another species of large, slow-moving fish. Some were curious, others annoyed, but most just went about their business without paying the loud, bubbly humans much attention.

There was a tap on my shoulder. Shep offered me his dive computer's instrument gauge, pointing at the reading: 70 FEET, 85 FAHRENHEIT. We wouldn't be getting hypothermia on this dive.

A stoplight parrotfish was grazing on a boulder up ahead. A bright blue, foot-long fish with multi-colored marking on its sides, it uses its round, beak-like mouth to scrape mouthfuls of coral. But it's not really the tiny coral animals parrotfish are after. They are herbivorous grazers, feeding on the algae growing on or within living or dead coral heads. The unmistakable scraping sound of their grazing can be heard from a great distance away. They excrete fine coral sand, vital to the formation of reefs and islands. The reef wall, and Calabash and Caulker too, are formed in part by the sandy excretions of parrotfish.

I passed over a small canyon, its sandy floor littered with living and dead coral heads. A few hundred yards away, Shelly was swimming up the ascending canyon floor following, I would later learn, a spotted eagle ray. I'd seen one myself a few days earlier off Caulker, a graceful brown creature that flies through the water on its wide, white-spotted wings. The larger manta ray—also a gentle plankton grazer—has a wingspan upwards of 20 feet; divers in the Cayman Islands occasionally hitch rides on the backs of their wings, an exhilarating service that some mantas appear not to mind providing. After a short time, the divers must let go for want of air or proximity to their boat, and the manta continues on its way, sifting the waters for copepods, fish larvae, and floating algae to power its beautiful flight.

A far more common sight are sting rays, timid creatures, 2 to 3 feet wide which, like the manta and eagle rays, are flattened cousins of sharks. Sting rays partly cover themselves to rest on the sandy bottom, much like flounder or halibut in northern waters. They feed on small fish and crustaceans by blanketing the bottom hiding spaces of their prey with their bodies, scaring them out of hiding by fanning sand and water with the undulating rim of their body, then swallowing them with their bottom-facing mouths. They have a sharp barbed tail to employ in defense. I once nearly collided with a sleeping sting ray while snorkeling in the barrier lagoon of Kosrae in Micronesia. The encounter gave us both quite a scare: first the sting ray, who exploded out of hiding, then myself, petrified by the sudden eruption of sand beneath and the undulating shape shooting off towards the reef crest. I've since paid much closer attention to the ground beneath me.

Sting rays like the sandy bottoms found on the landward side of the reef. In a healthy reef system, these shallows usually shelter great meadows of sea grass beginning just off the beach. These meadows appear uniform and unoccupied at first glance, territory to be transited on the way between the twin utopias of beach and reef, but they are vital for the preservation of both. They slow bottom currents, causing suspended sediments to fall and become trapped beneath their leaves; in this way the meadows keep the island and its beaches rooted to the seafloor and the lagoon clear for the corals.

Underwater meadows, like tide pools, reward the patient observer. Their long-leafed plants look terrestrial, and that's not a co-incidence. Sea grasses share with whales and sea turtles an evolutionary pathway from primordial sea to land, then back into the sea. The thick blades of grass are encrusted with dozens of species of sponges, flatworms, algae, and hydrozoan polyps, many of them quite beautiful. Sea-horses and immature fish hide among the swaying plants. In the Caribbean large conch shells are strewn about the meadows and sometimes, approaching slowly, I could see the eye stalks of the snail itself, scanning the bottom for tasty tidbits.

Sea cucumbers—animals that look very much like a discolored garden vegetable—appear in larger numbers, long, uniform strands of sandy excrement tracing their slow movements like streams of tooth-paste. These creatures vacuum the bottom of debris and sometimes harbor a tiny nocturnal pearlfish inside their bodies. Before daybreak, the pearlfish forces its way into the cucumber's anus tail-first, hiding there vampire-like until sunset, when it ventures out to feed.

Tiny garden eels poke their heads out of burrows but vanish in the blink of an eye when alarmed. I encountered one of these ener-getic jack-in-the-boxes in a sea-grass meadow near Belize's famous Blue Hole. I emptied my B.C. and settled belly-first on the sand a few inches from the eel's burrow. The tiny fellow popped out several times a minute to see if I'd decided to leave, vanishing in a little puff of sand if the answer was "no."

As the water deepens, the reef flat is littered with scattered colonies of coral, which grow denser as you approach the backreef slope. It was over such terrain that we'd been snorkeling a half hour before. Although the flat is protected, it's a challenging place for corals to grow. Storms and currents carry sand and sediments over the reef crest and spread them in the shallows. In fact, it is this process that forms coral cayes in the first place. A sand deposit can grow large enough to break the water's surface and, with a great deal of luck, vegetation becomes established, eventually consolidating an islet. But sand can also smother living corals. Sand, plus other sedi-ments and run-off from land, tend to concentrate in lagoon shallows, giving algae a competitive advantage over the corals. Very shallow water is often simply too warm for the corals to survive.

Corals are extremely sensitive communities. Most reef-builders thrive only in clear sea water with a year-round water temperature between 72 and 82 degrees Fahrenheit, and in depths of less than 230 feet. This is why they're absent from western continental coastlines, all of which are subject to colder currents. They can't compete with fast-growing algae in the nutrient-rich waters of upwelling areas, river mouths, or other sources of organic sediments. The ocean's richest ecosystems, coral reefs are found in its least nutritive coastal waters.

Coral reefs confounded ecologists for decades. Tropical waters are so clear since they are almost entirely plankton-free, and in any marine environment, phytoplankton are the producers, utilizing photosynthesis to capture the sun's energy to the benefit of every creature higher up the food chain. A typical ecosystem is often represented as a pyramid, with small numbers of carnivores supported by greater numbers of herbivores, themselves resting on an enormous base of producers. But coral reefs seemed to turn this pyramid on its head. Here scientists were confronted with a shocking mass of higher consumers—a profusion of fish of every imaginable appearance—and a slightly lower mass of snails, worms, crabs, and zooplankton-eating coral polyps. All of these consumers far outnumbered seaweeds and planktonic producers—an impossibility since the system apparently consumed more energy than it produced.

Part of the answer to this puzzle lay within the corals themselves.

Many people are surprised to learn that the colorful, rocklike boulders and reefs are the creation and domicile of tiny animals. The surface of a healthy coral head is actually a layer of coral polyps, an animal related to sea anemones. The polyps are tiny, from less than a millimeter to several centimeters in diameter depending on the species. They are cup-shaped creatures, the mouth at the top, surrounded by little tentacles, the rest of the body nestled in a limestone skeleton of the coral's own making. The tentacles wave in the sea, firing paralyzing darts at passing zooplankton and whisking the stunned prey into the polyp's mouth. Most species feed at night, closing their mouths for protection during the day. But this feeding activity only accounts for a small portion of their energy.

The remainder comes from microscopic algae called zooxanthellae that actually live within the tissues of the coral polyp. Zooxanthellae convert sunlight into energy via photosynthesis. The polyp assists this effort by supplying carbon dioxide, a waste product that most animals excrete but which these animals store for their algal tenants. In return, zooxanthellae donate their considerable excess food production in the form of carbohydrates. They are also thought to capture the calcium that polyps use to build their limestone shells. Each coral species is home to a distinct type of zooxanthellae, which give living reefs their surprising array of color. Neither polyp nor zooxanthellae can long survive without the other; corals are, in reality, plant-animals.

By exchanging materials in this fashion, corals can grow at a remarkable rate. They are colonial organisms, with each polyp attached to its neighbors in great sheets or branches. The polyps divide by budding, advancing the size of the colony by between 8 and 25 millimeters (0.3 to 1 inch) a year. A new daughter polyp quickly manufactures a limestone shell for itself, often growing on top of the dead skeletons of the previous generation. The shape of the colony reflects the way in which the polyps bud to copy themselves. Layer upon layer, year after year, the boulders and branches grow, driving the construction of the reef itself. The polyps reproduce sexually as well. On many reefs this occurs on the same night once a year, dozens of species releasing clouds of eggs and sperm into the sea, casting larval polyps to the currents to create new colonies.

Corals provide the building blocks for the reef, which are cemented together through the activities of other creatures. Worms, parrotfish, surgeonfish, and storms break corals and shellfish into sand, some of which washes over the reef to settle into the lagoon, often creating or sustaining coral cayes in the process. But a significant amount of sand is washed into the cracks, crevices, and open spaces of the reef itself. Here it is cemented into place by smooth, pink coralline algae, which encrust and consolidate the sand with their unusual limestone bodies. So hardy are these encrusting algae that they dominate areas of crashing surf where even the toughest corals cannot survive.

Nor can the reef thrive if divorced from its wider ecosystem. Zooxanthellae account for only part of the "missing" plant production that supports life on the reef. Many fish and invertebrates that make their home in the reef venture into nearby mangroves and sea-grass meadows to feed, often on the grasses and discarded mangrove leaves themselves. On Calabash, Shelly described how many fish and invertebrates ventured into sea-grass meadows and mangrove lagoons near the cay to feed, later returning to the reef to digest and excrete wastes; this daily commute feeds other creatures by adding sustainable quantities of nutrients to the reef system. Currents and storms also carry dead plant matter to the reef, where they are consumed by grazing fish, worms, or plankton.

The reef, lagoon, sea grass, and mangroves form one system. Take away one component and you compromise the health of all the others. And you may just lose the island in the process.

∞

SEVERAL FOOT-LONG YELLOWFIN SNAPPER were still following me and if I gestured towards one it would rush forward to see if I would offer it food. Even out here fish have learned about hand-outs. I'd become lost in the proliferation of fish: yellow angelfish, striped sergeant majors, damselfish and butterflyfish in assorted shapes, sizes, and colors, a reclusive 4-foot-long Nassau grouper, grazing parrotfish, fish with unusually large eyes, with whiskers, with false eyespots on their tails, or camouflaged to look like an algae-encrusted rock. It was difficult to concentrate on any one creature for more than a few moments before being distracted by another or by the demands of gauges, buoyancy, and navigation.

Each of the myriad fish species that flowed around me had its own unique ecological niche, often involving collaboration with other reef creatures. One often finds an inch-long red-and-white clown-fish amid the colorful, stinging arms of sea anemone, under which it lays its eggs or hides from predators. A clownfish must exercise great care and patience to take up residence in such a lethally armed home. It approaches with great care, fleetingly touching a single arm and withdrawing before being stung. Through many such lunges it slowly coats its body with the insulating mucus that the anemone secretes so that it will not sting itself. Elsewhere, gobies and wrasse establish "cleaning stations" where, with intricate body movements, they advertise their services to far larger predatory fish. These visit the cleaning station and remain docile while the cleaner fish remove parasites, loose skin, and fungi from their bodies, the insides of their gills, even deep within their mouths. Large-eyed squirrelfish feed on plankton, but do so at night to avoid competition.

Jonathan caught my attention, pointing to a boulder on the slope by which he was levitating. I drifted close to the coral head—a star coral of some sort, normally goldenrod brown. But a circle of bleached white coral had expanded from the center. Even here, in the deeper, well-circulated waters of the reef wall, some corals were being bleached, although nowhere near as many as in the shallow reef flat. A little farther along, Shep encountered a patch of partially bleached staghorn.

These corals were losing their zooxanthellae, thus their color, and without them their growth would slow or stop, the polyps becoming less resilient and more vulnerable to infection. If a coral head loses all of its zooxanthellae, the colony is likely to die. Many of the white boulders we'd seen on the reef flat were probably doomed. A wide variety of environmental stresses can trigger bleaching: drying (due to a particularly low tide); changes in salinity after, say, a heavy rainstorm; the ultraviolet radiation in intense sunlight; even oil spills. But the mass bleaching events that have made worldwide headlines since the mid-1980s usually occurred where sea water temperatures have been unusually high. The weather event known as El Niño displaces warm water currents and is associated with many widespread bleaching events. Corals may be an early warning system for climate change. Or perhaps they've simply faced too many human disruptions to cope with an additional natural stress.

I turned away from the whitening coral head. Fifteen feet below, a school of silver-blue fish hung nearly motionless against the reddish-brown cliff face, slowly turning their faces to the gentle wind of the passing current, moving like a child's mobile. We had started our slow ascent to the surface, careful to rise no faster than our smallest bubbles. So I continued watching the hanging school for several minutes before I realized why they were behaving with such lethargy. Drifting just underneath the school was a 5-foot-long barracuda, its long, prehistoric mouth ajar to reveal rows of pointy teeth, its chameleon-like body blending with the reddish-brown cliff. It seemed to be eyeing me, turning circles beneath the schools of fish and divers, trying to decide which would make a better lunch. Although we would have made an easy target—our skin soft and vulnerable—we were far larger than the prey size imprinted into the barracuda's brain. At the end of my Blue Hole dive, a surfacing diver dropped her mask, which sank 30 feet to the sea-grass meadow. Two sand-camouflaged barracuda shot out of nowhere and began lifting, prodding, tossing the small mask with their snouts. They also have the unnerving habit of circling divers.

The barracuda remained wracked with indecision. I'd reached 15 feet and stopped to give my cardiovascular system a chance to purge the build-up of dissolved gasses from my veins. I looked up at the shifting sky, hearing the quiet hum of Jerry's outboard motor waiting a safe distance away.

Time to return to the sweet air of our adopted home.

∞

ON OUR WAY BACK TO CAULKER, Mark took us on a side-trip to Swallow Caye, a swampy mangrove island whose shallows are home to a surviving clan of manatees. He approached the island slowly, cutting the engine as we passed warning buoys: "No engines—Manatees." Slowly, silently, he pushed us towards a large, unseen sink hole with a long wooden pole. We stopped in the muddy water 50 yards from the mangrove shore and waited in the tropical heat. The only sounds were the wind in the mangroves and the quiet dripping of water from Mark's pole, now laid across the gunwales.

Occasionally, a loud exhaling sound drew our attention to a cow-like snout poking briefly from the brown surface of the water. Slowly, over several minutes, the manatees approached the skiff with curiosity. Then, just under the bow a large shape slowly emerged from the gloom to reveal a huge, gentle, plumply rounded creature staring up at us with binocular eyes. So close, our faces only feet apart on opposing sides of the great divide. Then the manatee's snout broke the surface of the water, it took a deep breath of human-scented air, and then descended, ever-so-slowly, to return to grazing upon meadows of sea grass below.

∞

SCIENTISTS BELIEVE CORALS EVOLVED in the seas now surrounding East Asia, where the diversity and abundance of coral species is truly breathtaking. The Caribbean is home to approximately seventy-five species of reef-building corals, but the Indo-Pacific hosts more than three hundred species, many of which are found nowhere else, the product of 400 million years of evolution. Many may not survive the next century.

Of the estimated 230,000 square miles of coral reef in the world, about 10 percent is already dead or dying and another 30 percent is expected to decline significantly over the next fifteen years, according to estimates by Clive Wilkinson, a coral reef expert at the Australian Institute of Marine Science. People have damaged or destroyed reefs in ninety-three countries. Those in the western Pacific and Caribbean are believed to be worst affected.

Southeast Asia's reefs are in the worst shape of all. This part of the world has experienced rapid economic and population growth in recent decades. There are now 450 million people living in a region that relies on the sea for 60 percent of its animal protein needs. Development and hunger have conspired to undermine the foundations of the region's bountiful corals. Since 1945, half of the region's mangroves have been cut down to make way for cities, golf courses, shrimp farms, airports, factories, and shantytowns. Reefs are smothered in sediments and fertilizers, poisoned by oil and sewage, bombed with dynamite by desperate fishermen, and sprayed with cyanide by divers stunning reef fish for American aquariums and expensive live fish restaurants in Hong Kong. In the Philippines—

where blast- and cyanide-fishing were practiced extensively—nearly a third of the reefs are dead or severely deteriorated, and another 40 percent have less than 50 percent live coral cover. The figures are similar for Malaysia, Indonesia, and Singapore. Because of the continual increase in human populations and development pressures, six leading coral reef experts from the region predicted that, unless substantial actions are taken to protect them, "most of the coral reefs in the region will be exterminated" by 2033.

Nor is the situation in the Caribbean particularly rosy. The arc of island nations that delineate the eastern edge of the sea have all undergone rapid tourism-related development. Mangroves have been cleared, harbors dredged, artificial beaches dumped in inappropriate locations, and forests and ocean reef shallows replaced by airport runways. Haiti, Jamaica, Mexico, Panama, and other Caribbean nations have experienced rapid population growth, increasing pressure on reefs for food, tourist curios, even building materials.

Live coral cover along the north coast of Jamaica has fallen from about 60 percent in 1980 to less than 5 percent. Thick stands of kelp and seaweed have grown up where the corals once were, preventing recovery. The seaweeds are continually fed by sediments and nutrients from the densely populated island, and chronic overfishing has removed the plant-eating fish that might have kept them in check. Nature responded: To take advantage of the expanding fields of ungrazed kelp and seaweeds, the population of the algae-eating *Diadema* sea urchin exploded. These black spiny urchins lived in the reefs during the night, venturing out in large, slow-moving packs after dusk to feed on sea grasses and other algae. They were an amusing sight for night divers: dense packs of black pin-cushions venturing snail-like towards the sea-grass meadows. In 1983 these sea urchins suddenly succumbed to a disease that spread across the basin with ocean currents, reducing urchin populations by 99 percent. With nothing to keep them in check, kelp and other large algae overwhelmed Jamaica's reefs. Throughout the Caribbean corals are now barely holding their own against the kelps and weeds. All it will take is a random natural disaster to push them over the edge.

That's exactly what happened in Jamaica. The coup de grace came in the form of two hurricanes: Allen in 1980 and Gilbert in 1988. Reefs that appeared perfectly healthy before Gilbert never recovered from the hurricane's disruptions.

"Look, when you're in Belize and you see all these reefs in pretty good condition—you say 'yeah, there's extensive bleaching or coral damage inside the reef, but lots of fish and healthy-looking coral on the reef walls, looks all right.' But they may be right on the edge." Jeremy Jackson, an expert on coral reefs at California's Scripps Institution of Oceanography, had studied Caribbean reefs for decades and had just moved to La Jolla from Panama. When he talks about the region's reefs he sounds broken hearted. "The key lesson from Jamaica is that you can pile stresses on reefs and have them still look 'OK' until one day you add the straw that breaks the camel's back, and the whole system collapses. Before the collapse, Jamaica's reefs looked a lot like Belize's today." In hindsight it's obvious that overfishing and pollution were weakening the system. When the natural crises came, as they always do eventually, the reef was no longer able to cope.

During his professional career, Jackson has watched the deterioration of reefs off Florida, Panama, and the U.S. Virgin Islands. He's seen the shocking decline of staghorn and elkhorn coral across the region, the crash of the Jamaican reefs where much of the pioneering work on corals had taken place. "Jamaica's kind of a symbol since it was so beautiful before," he says. "There are a lot of depressed people like me out there in this field." In the space of a single professional lifetime, most Caribbean reef systems have visibly weakened, and many have completely succumbed.

Jackson started wondering just how much deterioration had occurred in the decades and centuries *before* he started his work. Using historic records, he pieced together how the reef communities might have looked when Columbus arrived in 1492.

"Think about the wildebeests and lions and all that on the plains of Africa," he said. "Well, there was a world out there in which the biomass of big animals amongst the reefs was greater than the biomass of the big mammals in the Serengeti plains. The absolute minimum number of 100-kilogram green [sea] turtles was something like 35 million. Think about that: 35 *million* 220-pound turtles grazing on crustaceans, sea grass, starfish, and mollusks. The productivity of those reefs must have been fantastic! The whole mind-set of scientists about what is a 'pristine' reef is completely wrong."

The turtles were wiped out by the middle of the eighteenth century, hunted to near-extinction to provide cheap food for slaves on sugar plantations. All large vertebrates in Caribbean reefs systems had been decimated by 1800 to feed sailors and slaves, and there has been subsistence overfishing of smaller vertebrates and invertebrates since 1850. The latest declines are merely the latest wave, the decimation of all but the smallest grazers to feed expanding populations. "Coral reefs," Jackson concluded, "aren't sustainable at even a tenth of current fishing rates."

Jackson and other scientists say that at a minimum perhaps 20 percent of coral reefs should be placed in strictly protected marine reserves, where they would be completely off limits. These reserves would act as storehouses of biodiversity and marine productivity, seeding other areas with life. Other areas would be afforded different levels of protection: no fishing here, only recreational uses there, this other place set aside for scientific studies. Healthy reefs would enhance the tourist industry, protect against erosion, and continue to move human visitors. Relieved of some chronic stresses, overall fish landings would probably increase significantly. But Jackson's not very optimistic.

"It didn't really hit me how hard it would be to change things until recently," he mused. "I was in the Philippines for a meeting of the International Coral Reef Initiative, and my hotel was right on the shore. I got up at 5 A.M. one morning and walked out to look at the ocean. The reef flat at this place was a half-kilometer wide and it was low tide. I saw all this strange movement out there. As the sun came up I realized that there were literally thousands of people out there in that reef flat, filling little bags with sea urchins, algae, a snail or two—anything they could find—and heading home.

"They were picking the reef clean for survival."

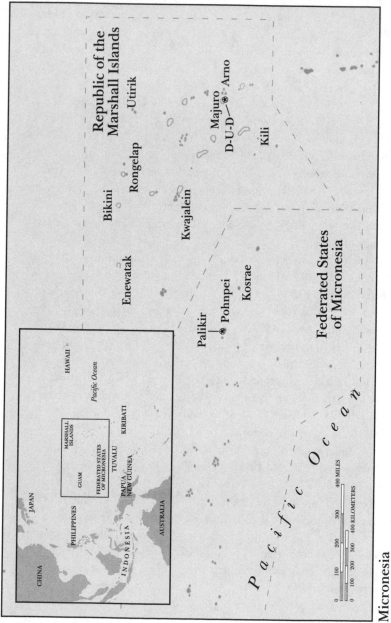

Micronesia

six
Paradise Lost

 THE PACIFIC OCEAN IS THE LARGEST geographical feature on Earth. It covers about a third of the Earth's surface, considerably more than all the continents, islands, lakes, and inland seas combined. At the equator it's 10,000 miles wide from the coast of Ecuador to the Asian mainland. The distance from its confluence with the Arctic Ocean to that with the Southern Ocean is more than 8,000 miles. To cross the Pacific—whether by canoe or wide-bodied jet—is to hurtle over a vast abyss punctuated by specks of solid ground, irregular constellations of tiny jewels set in unending celestial blue.

One such constellation encompasses the Republic of the Marshall Islands, a young nation in the very center of this vast ocean. There are over 1,100 islands in the Marshall chain, but if you added all of them together you'd have a parcel no bigger than the District of Columbia. But the Marshalls are not gathered together. They are spread over three-quarters of a million miles of the central Pacific, thousands of miles from any land mass of consequence. The Marshall Islands are remote even from one another.

"Islets" is probably a better word for these tiny, flat, fragile bits of land. Most of them are little more than glorified sand bars, anchored to coral reefs by stands of coconut and long-leafed pandanus trees, grasses, and shrubs. Few are more than a couple of hundred yards wide, and many are considerably narrower. Like the cayes of Belize, they are built of mounded coral sand produced on nearby reefs and piled up by lagoon and ocean currents. So they are low and flat. The average elevation of the Republic of the Marshall Islands is 7 feet above sea level.

These islets are gathered into twenty-nine ring-shaped chains, each surrounding a central lagoon, like little solar systems around a nourishing star. Each ring of islands and its associated lagoon is called an atoll—the offspring, if you will, of a liaison between volcanoes and coral polyps.

Millions of years ago a high volcanic cone stood where each atoll is today. Each cone spewed molten land around its base, building a high volcanic island, just as on the big island of Hawaii today. The warm shores of these young, mountainous islands proved excellent sites for corals, which formed fringing reefs along their shores. Over thousands of years, as the volcanoes went dormant and wind and rains gradually eroded the high volcanic island back into the sea, its shores contracted inward from their fringing reef of coral. In this way the fringing reef became a barrier reef surrounding the stubby vestiges of the original island and its ever-widening lagoon. Eventually the island was washed entirely into the lagoon, leaving only the reef to mark its long-lost shores.

The Marshallese people live on fragile islets that caught purchase on the atoll reef. From the air they look like strands of spaghetti turning on an unending surface of bright blue.

It's an unlikely setting for an apocalypse. But a century from now the Republic of the Marshall Islands may cease to exist. Theirs will be a most biblical fate. Created by fire, the Marshall Islands face a watery destruction.

❦

It had been a long trip, leap-frogging across the vast chasms of prairie, mountain, desert, and ocean separating airport arrival lounges in Virginia, Texas, and Hawaii. We landed at Johnston Island—halfway between Honolulu and Majuro—but weren't allowed off the plane. I stood in the door and watched thermals rising off the runway, causing the incinerator and chemical weapons storage bunkers to shimmer in the background. Barracks, garages, palm trees, and U.S. soldiers baked in the blinding tropical sun while the Army topped off the 727's fuel tanks. Nothing for a thousand miles in any direction, so someone in Washington decided it looked like a good place to dispose of nerve gas and other chemical goodies we'd been stockpiling. For the military mind, isolation has its own strategic value.

Three more hours of ocean and cloud passed below me. When Arno Atoll—the thinnest strand of sand and reef in all that blue—finally appeared, it looked like a lonely microbe on a microscope slide. Move the slide to the left and Majuro comes into view: first the densely packed settlement, then a thin strip of sand extending to a bulge of artificial landfill that supports the airport runway.

Dozens of sandy islets once perched along Majuro's living reef like an irregular pearl necklace. Now causeways connect half the islets, allowing one to drive 35 miles along the atoll, one of the longest drives in all of Micronesia. But there's literally no room for side trips. Majuro's islets are a few hundred yards wide at most, and often much less. Their average elevation is 7 feet.

Majuro Atoll, site of the Marshallese capital, is a beacon for youth from the outer atolls, drawn by junk food, beer, electricity, secondary education, and the hope of joining the cash economy. A third of the country's population now lives in Majuro's urban center, unromantically named D-U-D for Darrit-Uliga-Delap, three islets united long ago by landfill, potholed roads, and a massive fleet of imported taxi cabs. If you stand on the road, you often can see or hear the surf crashing simultaneously on the lagoon and ocean shores. On the startlingly narrow strip of sand between them is crammed three-quarters of the nation's infrastructure: the coconut plant and water distillery, the power plant and oil storage tanks, the port and warehouses, government offices, parliament, and embassies; car dealerships, supermarkets, gas stations, and stores selling imported goods; plus the densely packed houses and shacks of twenty thousand people. All this on 0.8 square miles of dry land.

Outside the terminal building the afternoon sun was blinding. Seconds after leaving the plane I was drenched in sweat. Even with sunglasses the entire landscape looked like an overexposed photograph. Reinforced concrete seawalls surrounded the building and runway, obscuring views of the shore. More seawalls flanked the potholed road leading to town: expensive imported granite boulders stacked in broad piles 8 and 10 feet high. The sputtering taxi crept along, slowing nearly to a stop before easing into enormous puddles or over unavoidable gaps in the roadway. Occasionally the sea closed in on both sides, held at bay by twin walls of piled boulders. To the right the ocean stretched to the horizon. On the left the atoll gently curved away to the left, boomeranging off into the distance across the 5-mile-wide lagoon. Of the more distant islands, only the tops of coconut palms could be seen, the land so low it was already concealed behind the curvature of the Earth.

Past the power plant and tank farm the town closed in on both sides: supermarkets and gas stations, foreign embassy compounds, the parliament building, the husk of a foreign-financed hotel never completed, densely packed homes with tin roofs, always with the ocean just behind in both directions, glimpsed through yards and alleyways.

Children everywhere: playing in yards, peering out of open windows, sitting in the road, chasing each other in parking lots, alleys, and walkways wherever open space could be found. When I passed them on the street, enthusiastic four and five year olds ran up to point at me yelling "*Belle! Belle!*" ("foreigner," literally "clothed person") in case I was confused about my identity. Later I learned to point back with the rejoinder "*driMazjal!*" ("Marshallese people"), which usually got squeals of laughter. With that, I could win over a significant portion of the country: half the population is under fifteen years of age. The national growth rate stands at 4.2 percent per year, and Majuro itself is growing at over 6 percent. Providing jobs, resources, education, and health care to this burgeoning new generation will be a staggering task.

I saw a muumuu-clad mother feed her baby a meal of Cheez-Whiz and Coca-Cola. I learned from Giff Johnson, the editor of the *Marshall Islands Journal*, that this is a fairly common practice. Although food is plentiful, significant numbers of infants show up at hospitals suffering from malnutrition. "Their parents feed them junk food almost exclusively," Johnson told me. "Adults will sell fresh vegetables or fruits they've grown for cash, then use it to buy processed foods from the States." They call it "the good stuff," but it's not doing them a lot of good. An estimated one-third of Majuro adults have diabetes.

Along a stretch of beach, a five-year-old boy was pooping in the surf. Disposable diapers floated in the lagoon behind him, part of a growing heap of imported trash nobody can dispose of. On Majuro, there is no land for landfill, so the ocean shore of D-U-D is one long pile of aluminum cans and tires, rusting husks of automobiles, refrigerators and stoves, heavy machinery and plastic packaging. Once, island wastes were biodegradable and could be left on the shore to rot and wash out to sea. But not the detritus of Western civilization. Some of this more resilient garbage is collected and trucked out to a stinking trash heap to build new land from scratch. The government hopes that Majuro can replace its eroding soil with great mounds of rubbish.

Container loads arrive on Majuro from California and Guam: used cars, plastic cups, cases of Coke and Budweiser, and boxes of plastic shopping bags. The ships leave empty since the Marshall Islands don't produce much of anything. The Marshallese pay for these goods with U.S. greenbacks, just as they pay for their imported fuel, spare parts, construction materials, frozen meats, canned vegetables, clothing, and air conditioners. Ultimately, most of this money comes from annual U.S. grants paid to this former U.S.-administered trust territory as part of a 1986 independence agreement. America captured these islands from the Japanese during World War II and used them as a military supply base and atomic testing range. By the late 1970s, however, colonialism had become a dirty word and the Marshallese wanted control of their own destiny. They worked out a deal: The United States granted the nation its independence but rented its military sovereignty and the right to operate the Kwajalein Army Missile Range in exchange for roughly $1 billion paid over fifteen years. With this money, the Marshallese were to build a modern, self-sufficient economy, while providing for those citizens who had lost their homeland, relatives, or health during the U.S. atomic testing program.

Today the U.S. cash transfers aren't just critical to the Marshallese economy, they *are* the economy. They account for two-thirds of the country's Gross Domestic Product. The Marshallese have built a way of life around the magical arrival of the annual grants. They receive other support as well. Mail is delivered by the U.S. Postal Service, disasters are attended to by the Federal Emergency Management Agency, and post-secondary education is built upon Pell Grant funding. The government has borrowed heavily against future grant payments and squandered the money on white elephant projects.

The money runs out in 2001. That, however, may be the least of the country's worries.

∞

"IT WAS A CLEAR, SUNNY DAY. No wind, no clouds in the sky, no warning at all," Abraham Hicking began. "I was a schoolteacher at the time at the elementary school in Darrit. It was 8:30 in the morning and we were in the middle of a lesson when the wave hit. We just stopped class and the children gathered at the window to see."

Outside the window the intruding ocean quickly rose up to the knees of shocked pedestrians. The water inundated the entire neighborhood, spreading clear across the island and on across the lagoon. Currents pushed wooden houses, refrigerators, and automobiles into the sea. Darrit remained swamped for most of the day, with living rooms underwater and squealing pigs swimming around in the lagoon.

"When the sea rises," Hicking said, "there's nothing to stop it from going right over the island."

To this day Majurans don't know what caused the freak 1979 wave surge. Maybe it was produced by an undersea earthquake thousands of miles away, or perhaps by a typhoon. Under clear, sunny skies people found themselves suddenly dragged into the sea, treading water alongside bewildered pigs and other livestock. By most accounts there were no deaths owing to the balmy conditions, but 144 homes were destroyed. President Jimmy Carter declared the island—then a U.S. trusteeship—a national disaster area.

"People have a gut feeling that the sea is changing, that there have been more wave surges and more erosion than we have experienced before," continued Hicking, now middle-aged and the chief of water quality monitoring at the national Environmental Protection Agency in Majuro. We sat in his tiny office, shared with Steve Eke, an Australian volunteer, and located only a few feet above the shore of Majuro's vast deepwater lagoon. The whole EPA building was no larger than a trailer home; a 20-foot wave would have little trouble pushing it off its sandy moorings and into the sea. "One thing is clear," he said. "With sea-level rise there is nowhere to retreat."

"It's the combination of sea-rise and freak events that are going to be the biggest problem," Steve added. "Instead of having a one in a thousand chance of a freak wave like the one in 1979, it might be one in a hundred because of climate change and a rise in sea-levels. Storms will take away more land and there will be more of them. If sea-level rise predictions are right, there won't be much left of this country."

Hicking smiled gently as if to say, "such is life."

This was a disposition I'd encountered in many Pacific islanders: extreme calm and passivity in the face of adversity. "I could run out into the street and take a baseball bat to that passing car," an embittered American expatriate told me, "and they'd just shrug and walk on home. Just once I wish they'd get angry!"

So I pressed Hicking on how he felt. He was born in Kiribati—another nation of atolls to the south—but had lived on Majuro for much of his life, married a Marshallese woman and had two children, now nine and seventeen. Greenhouse gas emissions from rich, faraway lands might wipe both Kiribati and the Marshalls off the map. Didn't that bother him?

He thought for a moment. "When I fish in the lagoon I wonder if the coconut trees I see on the shore will be there two decades from now. I wonder if my children will have a home to come back to, or what sort of home it will be. It's their generation that will really have to confront the sea-rise issue, but they all want to leave the country anyway. My children want to live the U.S. lifestyle. I think it's the infrastructure that attracts them—the theaters, roads, and shopping malls. Parents want to have their children educated abroad and working overseas. Everyone looks away from here. Only those who are trapped behind will be forced to face the sea level problem.

"The fear is here, but reality is elsewhere," he concluded. He wore the same gentle smile he'd started with.

တ

CLAUDE HEINE USED TO WORK at the Robert Reimers marine yard, but now he just hangs out there. He walks with a limp, the result of a disabling accident years ago. He sits around the dock whiling away the days with beer, conversation, and the occasional consumption of the freshest fish available anywhere. Today it's a large rainbow runner caught by a friend just offshore. The bowline of his skiff was not yet tied to the dock when the fish was handed ashore and Claude began cutting delicate strips out of its side. Bottles of soy and Tabasco were produced, their contents dashed onto the raw, red flesh and dropped into the mouth in one great coil. One was handed to me in such a kind and matter-of-fact way as to be difficult to refuse. I'd preferred my fish cooked, but I was to learn the error of my ways. Looking the animal straight in the eye, I dropped a piece of its flank into my mouth, bracing for the worst. More delicious and delicate than can be described: sashimi, fresher than that of Tokyo's finest restaurants.

Claude talked about this and that: politics (corrupt), fishing (worse than before), women (assorted), family (large). Eventually the conversation turned to erosion, to falling trees and vanishing beaches. "Oh, we've made a real mess of it," he said, opening another can of Budweiser. "You know what the priest says in church? It's the flood that's coming, the great flood like in Genesis. Oh boy, are we ever going to pay for our sins." With that he broke into hearty laughter, but his eyes showed only a sodden sadness.

⟐

SIGNS OF EROSION ARE EVERYWHERE. The beach near my rented apartment in Darrit is vanishing, taking with it backhoe-sized chunks of precious topsoil. The island road is eroded and, worst of all, the palm trees lining the top of the beach are dead or dying, the sea having exposed their roots; fallen palms are a common sight all along the shore. In front of the weather station a section of the concrete sea wall a hundred yards long has been battered to pieces, and the sea has started consuming the adjacent lawn, one of the few open areas in all of D-U-D.

Out on Laura—the island at the far end of the causeway, named for Lauren Bacall by homesick U.S. servicemen in World War II—a favorite family picnic spot has melted into the lagoon. Once a wide beach stood here, with great stands of coconut palms under which people would eat great Marshallese feasts of imported junk food during the hottest part of the day. Now the sea has carried away half the beach.

The shacks are packed tightly along the ocean shore in D-U-D; pig crates and flocks of free-range chickens fill in the gaps. Many homes sit right on the shoreline. So does the local cemetery, its tombstones looking out across the Pacific. Every so often a storm surge inundates the burial ground, carrying away soil, exposing coffins and uprooting headstones. Families come, rebury their dead, and replace lost headstones. On this particular day they rest in peace, but piles of flotsam litter the sandy yard. A piece of a plastic bucket is stuck against one headstone, the remains of nylon fishing lines twist around the top of another. Pieces of Styrofoam collect where flowers normally would be placed. Mounds of dried seaweed are everywhere.

There aren't any streams, rivers, or lakes in the Marshall Islands—there's not enough land to form a watershed. Rainwater simply seeps into the sandy soil until it reaches the salty ocean water that permeates the atoll's coral foundations. Because fresh water is less dense than sea water, it sits on top, forming a "lens" of trapped water that is tapped by coconut trees, garden crops, and Majuro's exploding human population.

Here, as elsewhere in the Pacific, this precious freshwater lens is becoming inundated with briny ocean water. On Majuro, this may be because the population is exhausting the lens faster than it can regenerate. Or it may be due to the mysterious absence of seasonal rains in recent years, or because the erosion of land is compromising island geology. Whatever the reason, an island without water is an island without agriculture, an island with fewer plants and trees to fight erosion, an island less fit for human habitation.

∽

THE WORLD'S SEAS ARE RISING. That much we know for sure. They've been slowly creeping upwards since the last ice age ended, eight thousand years ago. Since then they've risen 120 meters, as the massive glaciers that once encased Canada, the northern United States, Siberia, and northern Europe slowly melted. The geologic record suggests that when the continental ice sheets were breaking up, world sea levels rose by 2 or 3 centimeters per year, but in recent millennia the rate of ascent gradually declined to about a millimeter per year. Then, starting in the early nineteenth century, the rate began to accelerate. Over the last century the world's oceans have risen by 10 to 25 centimeters (3.9 to 9.8 inches), significantly faster than the average rate of sea-rise for the past five thousand years.

Most scientists who study the issue believe that two recent trends—increasing carbon dioxide (CO_2) levels from the burning of fossil fuels and accelerated sea-level rise—are closely related.

Carbon dioxide is a greenhouse gas. Like the glass in a greenhouse, CO_2 allows light to pass through to the Earth's surface but prevents the resulting heat from escaping. If the CO_2 concentrations increase, more of the sun's radiation is trapped in the lower atmosphere as heat, raising surface temperatures in the process.

Weather records show that average surface temperatures on the planet have increased by between 0.3 and 0.6 degrees Celsius

(0.5 degrees to 1.0 degrees F) since the late nineteenth century. Global temperatures are projected to increase by between 1 and 3.5 degrees Celsius (1.8 degrees to 6.3 degrees F) by 2100, with a best estimate of 2 to 3 degrees. This represents the fastest warming rate in ten thousand years.

Global warming affects sea levels in three ways. First, a warmer atmosphere will eventually warm the oceans as well. Like most substances, sea water expands when heated. Computer models suggest that thermal expansion alone will raise global sea levels by 28 centimeters, or almost a foot.

Second, warmer temperatures are melting mountain glaciers worldwide. Incredible quantities of meltwater are being released into streams and rivers, eventually running to the sea. This process is thought to have raised the ocean level by 5 centimeters over the past century. In the next century glaciers will retreat rapidly, raising the waters by about 16 centimeters, or half a foot.

Finally—and this is the least understood part—there are the polar ice caps. Climate models predict, apparently correctly, that global warming will be most severe at high latitudes. Further melting of Greenland's massive glaciers will bring a 6-centimeter rise in sea level over the next century. But scientists assume a zero or even negative contribution from Antarctica, since increased snow fall over permanently frozen regions should more than offset water released from melting glaciers or ice shelves. There is, however, a massive, precariously balanced store of frozen water in West Antarctica that is extremely vulnerable to climate change. For two decades scientists have been worrying over the possible break-up of this ice sheet—a catastrophe that would result in an additional *18 feet* of sea-level rise. More on that in chapter 7.

Setting aside the third scenario for now, models expect thermal expansion and the melting of glaciers and ice sheets to cause a total rise in sea levels of between 20 and 86 centimeters—or about 1 to 3 feet—with a best estimate of about 50 centimeters (20 inches). These are conservative estimates, the product of careful, comprehensive assessments by hundreds of scientists from around the world. (All the figures in this section come from the current climate change assessments of the Intergovernmental Panel on Climate Change [the IPCC], an international scientific working body created by the world's governments under the auspices of the World Meteorological Organization and the United Nations Environment Program. The IPCC is charged with providing authoritative international statements of the current scientific understanding of climate change and its likely impacts. Three working groups integrate the ongoing work of several hundred climatologists, glaciologists, oceanographers, geologists, atmospheric chemists, and geophysicists. Coordinated data are fed into different types of complex computer models, models which improve with advances in technology and scientific knowledge. It's the IPCC's predictions that prompt the debates one hears about climate change, global warming, and the possible destruction of the Republic of the Marshall Islands.)

One to three feet may not seem like much, but such a change would produce catastrophic results. Some very low-lying areas would become permanently inundated by the rising waters. But the real danger comes from the combination of higher water and extreme weather events such as hurricanes, tsunamis, and floods. Disaster planners prepare for such events by examining the weather record, calculating the average frequency with which a weather event of a given intensity can be expected to visit a particular area: a "fifty-year storm" is the worst weather event a particular area would encounter in a statistically typical fifty-year period. Combined with sea-rise, even a two- or five-year storm might inundate and erode vast low-land regions around the world, poisoning cropland and water supplies with salts, carrying away homes, infrastructure, and soil, even permanently removing the land itself. As the seas rise, most land loss will occur during such weather events.

To make matters worse, global warming is expected to change weather patterns. Hurricanes and typhoons might regularly strike regions that are rarely affected by them today. Changing rainfall patterns will bring increased drought or flooding, making land more vulnerable to attack from the seas.

Even without taking increased storm activity into account, a 3-foot rise in sea level will result in a cataclysm unknown in recorded history. According to the IPCC's 500-page study on this topic, one-sixth of Bangladesh would be inundated, displacing 15 million people. Eight million Egyptians would lose their homes in the Lower Nile Delta; the ancient city of Alexandria would be evacuated. Nearly 4 million would flee low-lying portions of Nigeria, where much of that country's vital petroleum industry is located. Seven million will face inundation in India, two million in Indonesia. Gambia will loose its capital city. These very poor countries will lose fertile cropland, freshwater supplies, towns, ports, and even cities while gaining millions of refugees. The political, social, and economic consequences can only be guessed at, but I suspect they will not be pleasant.

Wealthy nations may be able to buy themselves out of the problem, but at great expense. Regions that would be inundated under the three-foot scenario include much of south Florida; New Orleans and the Gulf Coast of Louisiana; Galveston, Texas; Charlottetown on Prince Edward Island; and swaths of New York and Boston. The cumulative cost of protecting U.S. and Canadian coasts and cities from such a rise in sea level is generally estimated at between $40 and $400 billion over the next century. Europeans must protect or lose the Dutch and German North Sea coasts, the Rhone and Po River deltas, and Venice. On both continents, levees and seawalls will not save many existing beaches, wetlands, and near-coastal aquifers.

For many small-island states, sea-level rise will not be simply expensive but apocalyptic, a most biblical end to their land, culture, and history. A 3-foot rise would consume much, but not all, of the Federated States of Micronesia and the Solomon Islands in the Pacific and the Maldives in the Indian Ocean. It would completely submerge at least three nations that consist entirely of atolls: the Marshalls, Kiribati, and Tuvalu.

In 1992, the U.S. National Oceanic and Atmospheric Administration (NOAA) funded a detailed study of the effects of predicted sea-level rise on Majuro Atoll. It's sobering reading, particularly if you're Marshallese. Even at its current "normal" rate, sea-rise will flood 10 percent of Majuro's dry land over the next century. A 3-foot rise will eliminate the atoll entirely.

The study examines the country's options—in public policy lingo, its "coping strategies." One is to protect the shore with an enormous armor of stone and tribar structures. This is technically possible if sea-level rise is modest, but the costs are prohibitive.

D-U-D proper, the built-up area at the west end of Majuro, is only 0.8 square miles. It could be protected from the effects of a 1-foot sea-level rise during a fifty-year storm event at a cost of $46 to $52 million—more than half the nation's annual GDP, including its U.S. grants. To protect D-U-D against a direct hit by a typhoon under such conditions would cost $100 to $105 million, too much for the Marshallese economy to bear without foreign assistance.

Saving all of Majuro Atoll (3.5 square miles) from this "modest-rise and fifty-year event" scenario would likely cost many hundreds of millions. Saving the entire country (70 square miles) would require the entire output of the Marshallese economy over at least a century. These are estimates for the "modest" scenario.

The NOAA study doesn't even attempt cost estimates for defending Majuro against the combined effects of 3 feet of sea-rise and a fifty-year storm; it would be prohibitively expensive and perhaps physically impossible. So if sea-level rise is serious, other strategies will have to be explored: accommodation, retreat, or abandonment. Unfortunately, on an atoll these all mean about the same thing. "Retreat may take the form of Marshallese moving to least vulnerable areas within or among atolls in the country, with Majuro Atoll being developed as the ultimate safe haven for the nation," the study reads. "Full retreat of the entire population of Majuro Atoll and the Marshall Islands must be considered in planning for worst case [sea-rise] and climate change scenarios."

In other words, the Marshallese may be forced to leave their native land, never to return.

❧

IT'S GENERALLY THOUGHT that the Marshallese colonized these atolls about two thousand years ago. Like other Pacific peoples, they are descendants of ancient ocean navigators who routinely crossed thousands of miles of open water in outrigger canoes long before the Greeks found the Black Sea. They accomplished this entirely through the careful observance of natural signs: the shape and direction of ocean swells, the presence and flight paths of various types of seabirds, the wind and, of course, the stars. They spread out to populate a quarter of the Earth's surface, from Easter Island to Hawaii, the New Hebrides to New Zealand. The Marshallese have lived here far longer than Slavs have lived in the Balkans or Anglo-Saxons in the British Isles.

Now they are poised to join a new class of landless, environmental refugees. If serious sea-rise conditions come about, the Marshallese will probably find themselves unwanted guests in some far-off foreign sanctuary, maybe a new American Indian reservation. If Marshallese culture is separated from the conditions in which it grew, it will rapidly erode. Within a generation or two their language will probably be lost. Their leaders will struggle to seek from the industrial world proper medical and social services, trust funds to support future generations, and compensation for the loss of a homeland. Their more enterprising descendants will no doubt assimilate, but the Marshallese people will have ceased to exist.

For the indigenous residents of four Marshallese atolls, this grim situation will seem all too familiar. The people of Rongelap and Bikini already know what to expect if the nation must climb into boats and face a life of exile. They have been living in forced exile for decades, ever since the outside world rendered their atolls uninhabitable.

❧

EARLY ONE MORNING IN 1954 an eleven-year-old boy named Norio Kebenli was preparing to fish in the lagoon of Rongelap Atoll in the northern Marshalls. The sun was dawning, as it always does, in the east.

Then a second sun rose in the west. Kebenli's life, like those of all Rongelapese, would never be the same.

"The light appeared in the western sky and became bigger and bigger," he recalled forty-four years later in the Rongelap municipal offices, now located on Majuro. "It was a huge cloud with a yellow and orange, mushroom-shaped, and the light was so strong it hurt my eyes. It filled the whole western sky." A few hours later it "snowed" for the first time on Rongelap: From the sky fell thick flakes of a strange gray ash that curious children played in and old people rubbed on their bodies in the hopes it would have curative powers.

The United States had just exploded its largest thermonuclear device ever over Bikini Atoll, 125 miles over the western horizon. The Bravo test—the largest nuclear bomb in U.S. history—exploded over Bikini at dawn, March 1, 1954, with the strength of 15 million tons of TNT, the equivalent to 750 Hiroshima bombs. The blast vaporized the test island, eradicated parts of two adjacent islets, and punched a mile-wide crater in the reef. The fireball could be seen throughout the Marshalls.

Bravo was not the first or last nuclear weapon detonated on the Marshall Islands. Between 1946 and 1958 the United States exploded sixty-seven atomic and thermonuclear weapons on Bikini and Enewetak, two atolls in the northern Marshalls. Their inhabitants were evacuated long before the blasts; the Bikinians remain exiles to this day. With one exception, the prevailing winds are believed to have carried the fallout out to sea, away from the populated atolls of the northern Marshalls.

The Bravo test was the exception.

The Atomic Energy Commission (AEC) would later claim that a last-minute wind shift took test personnel by surprise and spread fallout westward across Rongelap, Utirik, and Ailinginae atolls. Decades later, when the appropriate documents were declassified, the truth came out: The wind had shifted in this direction seventy-two hours before the test. AEC officials decided to go ahead as scheduled.

The morning after he witnessed the rising of a second sun, Norio remembers that people began bleeding, vomiting, and losing their hair. He and the other eighty-four islanders were ushered on board a U.S. destroyer that had arrived overnight, and the Rongelapese, like the Bikinians before them, became people in exile. They were moved around from one island to another. U.S. government doctors inspected them several times a year but usually told them little or nothing. Young children were occasionally whisked away to Hawaii to have their thyroid glands removed. Several severely deformed "jelly babies" were born. Cancers of all sorts developed over time.

James Matayoshi, mayor of Rongelap when I visited in 1998, hadn't been born yet on the "day of the two suns," but his brother was two years old. Matayoshi stands head-and-shoulders shorter than his older brother, the result of thyroid problems. Many older Rongelapese have scars on the sides of their necks from having their thyroids removed.

In 1957 the Rongelapese were sent home and, eleven years later, President Johnson announced that it was safe for the Bikinians to return. On both islands, houses were built and the people went back, fished in the lagoon, harvested coconuts, breadfruit, and taro, feasted on crabs, and began rebuilding their lives. Sadly, the American scientists were mistaken. Although background radiation had dropped to normal levels, Cesium-137 had become concentrated in the soil, in reef fish and coconut crabs, in plants and fruit, and eventually it lodged in the flesh and bones of the people themselves. Doctors examining the Bikinians finally ordered a new evacuation in 1978. But examiners at Rongelap somehow failed to notice the mounting evidence of radiation poisoning among their patients.

"I personally feel that my people were being used as guinea pigs," says Matayoshi, whose desk is piled high with declassified U.S. documents referring to "test subjects," "control groups," and how valuable the collected data were because Marshallese are "more similar" to Americans than lab mice. "These kinds of quotes are devastating. I mean, they knew the wind had shifted, they knew we'd be irradiated, and they had to have known we were getting sick after they sent us back. . . . What sort of people do this to human beings?"

By the early 1980s, the people knew something was very wrong and repeatedly asked to be evacuated. the United States said it was unnecessary, and the Marshall Islands government failed to act. In the end the islanders were evacuated by Greenpeace. The Rongelapese and Bikinians remain in exile to this day.

∾

THESE DAYS MOST BIKINIANS LIVE on one of two islands. Nine hundred and fifty people kill time on Kili, a remote, windswept, 200-acre island that isn't even part of an atoll. There's no lagoon, so nobody can fish. They just wait for the cargo ship to arrive with soda, processed snacks, frozen meats, and other goodies provided by Uncle Sam. Another 250 Bikinians live on Ejit, a few islets up the Majuro chain from D-U-D. They live in relative tranquillity—there are no cars, roads, or supermarkets here—but the hustle and bustle of D-U-D is only a short boat ride away.

I made the twenty-minute trip to Ejit in an outboard-powered skiff with Jack Niedenthal, the Bikinians' Pennsylvania-born Trust Liaison and external relations officer. Jack, now forty, had come to the Marshalls fourteen years earlier as a Peace Corps volunteer on the outer islands and never left. A lot of people fall in love with these islands. I met Peace Corps alumni everywhere: the hotel owner, the Nuclear Claims Tribunal spokesman, the newspaper editor, the lawyer, and the schoolteacher. Most came during the Vietnam War and found it a lot saner than home. Jack, who came much later, found the Marshallese to be "the nicest people I'd ever met." Now he's fluent in Marshallese and is married to a Bikinian woman with whom he has four young children. It's a matrilineal society, so by marriage Jack shares title to his family's ancestral land on Bikini. In a sense, Jack's become a Bikinian.

A few minutes from the dusty streets and beachfront trash piles of D-U-D, we entered another world. On our right, little islets scrolled by in quick succession, tiny blobs of sand held down by coconut or pandanus trees. Ahead the arc of islands vanished into the horizon; of the furthest only the tops of the coconut trees could be seen, the land hidden behind the horizon. Astern was a mirror image but with the buildings and infrastructure of D-U-D dominating the middle distance. Above, the equatorial sun burned with unrelenting intensity.

We waded onto Ejit's lagoonside beach and, flip-flops replaced, wandered on sandy ground between modest, well-maintained bungalows. Jack was taking me to call on Kelen Joash, a kind, gracious, sixty-five year old and one of the few Bikinians who can still remember the 1946 evacuation. We sat around his doorstep with his five-year-old granddaughter, a communally owned dog, and a central air conditioning unit while Kelen simply and patiently described the community's long ordeal: the painful evacuation from Bikini, near-starvation on the remote island where the Navy first dumped them, mass moves to Kili, from Kili to Bikini, Bikini to Ejit, health problems, doctors, lawsuits, unanswered questions. Jack deftly rendered Kelen's melodic Marshallese into laid-back Sixties speak.

"Bikini is wonderful," Kelen said. "It's got everything. It's got bird islands. It's got fish. It's got turtles. It's got every single kind of food you can imagine and it's plentiful. It's not a place where you go hungry. It's a beautiful place where you can just sail around and have a good time. There's breadfruit, bananas, pumpkins, and coconuts, but I don't think we're supposed to eat them now. The breadfruit and bananas are ripe on the tree and the first thing we want to do is eat them, but they [the Department of Energy personnel on the island] say, 'no you can't eat them right now.' And we're like, 'Why did you plant them if you can't eat them?' Its very hard for us to understand this."

I asked if he was angry at the United States for taking their island, neglecting his people, and mistakenly sending them back to a contaminated land. "Anger with the United States isn't a good way to put it. I mean, look at the United States. America is the most powerful country in the world, so we have to follow what they do or say to us. It's not like we ever felt like we had a lot of choices to make. The way I look at it, it was a time cf war when they came to us. The world was at war and, well, they did ask us whether or not we would leave. And they told us if the island didn't turn to glass after the bomb we would be able to go back. To me it was just the way the world was at the time. People were doing things like that."

Now his community was trying to decide whether they should trust the United States a second time and return to Bikini. Radiological studies show high levels of cesium contamination in soil and vegetation, but normal background radiation levels. If local foods are eaten sparingly and the soil in residential areas is scraped and replaced, the experts say the islanders could safely return. Getting the money to do so may take many more years.

Which brings us to the ultimate irony. After half a century in exile, the Bikinians may return to their homes just in time to see them vanish under the waves. A tiny traditional community in one of the remotest places on Earth will have borne the fury of the industrial world's two greatest demons—the Bomb and climate change—neither of which they played any part in creating.

I'd asked Jack about this earlier. He argues that the community should return, and goes there with his eight-year-old son twice a year so he can appreciate what is there for them. Of climate change, he shrugged: "That's not something I see any point in worrying about."

Kelen said he was concerned about the disappearance of islands, not in the future but in the past. "Some of the islands were vaporized [by the Bravo blast] and those places where they were are now part of the sky. Those were once bird- and oyster-collecting islands and now there's nothing there and it makes me feel a little sad." Then his voice sounded angry for the only time during our talk. "What are the people of America going to give us for those islands? We can never get them back."

What will the industrialized world give the Marshallese in exchange for the loss of their nation?

∞

SOME ARGUE THAT THE EROSION OF MAJURO may not be due to sea-rise at all. Population and development pressures may be doing the job on their own. Islanders have been mining sand from the lagoon to make cement for Majuro's myriad construction projects; unfortunately, the dredging close to shore may have caused the sloughing of nearby land. The construction projects themselves compromise the integrity of the island by stripping the landscape of trees and vegetation.

To protect homes and buildings, property owners construct sea walls from cinder blocks and stone boulders, or by heaping old automobiles and heavy equipment on the shore. These may protect their land from inundation, but they don't solve the erosion problem. The currents and wave surges are simply deflected to somebody else's land, sometimes with greater strength. That landowner builds a larger sea wall, displacing the problem again. This can trigger a "wall race" of sorts, as property owners and the government build and strengthen seawalls to out-do their neighbors. This competition might seem to explain away the Pacific nations' assertions that sea-level rise is already occurring.

Only the problem isn't confined to the densely populated main islands of Majuro, Kwajalein, or Tarawa atolls. Erosion, saltwater poisoning of crops and freshwater lenses, and property damage also plague sparsely populated islands throughout the central Pacific.

At the time of my visit to the Marshalls, nearly two decades had elapsed since anyone had lived on Rongelap. There are no roads or seawalls, no sand dredging or clear-cutting, no man-made causes of shoreline erosion. Apart from the radioactive isotopes embedded in the soil, flora, and fauna, the islands are pristine.

Like the Bikinians, Rongelapese have returned home for short visits in recent years. There are decisions to be made: Should we return to the atoll? Is there a future for our children there? Is a better life waiting for us there? The majority of Rongelapese have never lived on the atoll. For the rest it is becoming a distant memory. The trips help all decide what their hearts want to do.

But Rongelap Mayor Matayoshi says the elders have noticed things that younger generations miss. "They say the coastlines have changed. Where the cemeteries once were there is just a lot of sand. The coastline has shifted, and they see the same sort of erosion problems that we have here in Majuro."

"As a Marshallese citizen I'm really worried," he continued. "If this climate change is really happening, where is the future of the Marshall Islands?"

∞

NOT SURPRISINGLY, the governments of small island states attach greater urgency to climate change issues than most countries.

In December 1997, high-level representatives of virtually all of the world's governments met in the ancient Japanese capital of Kyoto to take coordinated action on global warming. They argued and postured and ultimately came up with an extremely modest agreement to curtail greenhouse gas emissions. By 2010, the United States would reduce such emissions by 7 percent below 1990 levels. Europe would reduce them by 8 percent, Japan by 6 percent, and the developing world not at all. Organizers called the targets "binding," but in fact the agreement contained no enforcement measures, and any nation could opt out with one year's notice.

If the United States and other nations actually sign and ratify the Kyoto treaty it will be a step in the right direction. It would be the first time governments have considered restraining greenhouse gas emissions and would require an improvement in energy efficiency similar to that which followed the 1970s oil shortages. Climate change aside, that will make our economy stronger, less wasteful, and more efficient. If the scientists are right, it gives us a head start on the more dramatic energy revolution we will need to stabilize our planet's climate before it crushes our fragile, nascent global civilization. Kyoto represents a cautious, wait-and-see approach; by 2010 the evidence for climate change should be much clearer and then, if necessary, we'll take bolder steps to save ourselves. But by then it will probably be too late for the Marshalls, Tuvalu, or Kiribati.

"Sea-level rise is an immediate problem which needs to be dealt with now," Espen Ronneberg was telling me on the line from New York. He had recently returned from Kyoto, where he was both the Marshallese representative and the conference vice president. He had spearheaded the efforts of small island countries to call attention to the immediacy of the threat climate change poses to their continued existence. But small countries have small voices on the world stage; their concerns weren't shared by more influential nations. "We always seem to get the reaction from the U.S.: that this is something we can put off, study, and see how it goes," Ronneberg said. "We don't really have that luxury."

This difference in perspective continued to crop up at Kyoto and during subsequent meetings on climate change. Developed nations—the United States in particular—acted on short-term interests, watering down targets for greenhouse gas emission reductions while pushing for unlimited "emissions trading." (This provision allows developed countries to avoid real emissions reductions by paying developing nations to plant or preserve forests as "carbon sinks.") Meanwhile, China and India refused to undertake any binding commitments at Kyoto—a key U.S. demand. Leaders of small island nations watched with despair as national self-interest and short-term thinking eroded draft treaty provisions even faster than the sea was taking away their homelands.

They did their best to organize, with some success. Forty-seven small island states came together to form a diplomatic block for the climate change negotiations, the Association of Small Island States or AOSIS. They passed resolutions, held press conferences, and sent their leaders to meeting podiums to press the only argument they had: the moral one.

Following are excerpts from a typical speech, this one given by Bikenibeu Paeniu, the prime minister of the atoll nation of Tuvalu, at a November 1998 meeting in Buenos Aires of the parties to the Kyoto climate change treaty. The delegates had gathered to follow up on agreements made at Kyoto, but bickering prevailed over action.

"We are the most vulnerable of the most 'vulnerable' countries. I need not remind the Conference that the whole issue of climate change to us is not [one of] economic and politics, but is of life and death," Paeniu said.

He described how Tuvalu was already spending millions of dollars to respond to devastation caused by the adverse effects of climate change, money that would otherwise have been spent on education, health care, and the reduction of infant mortality. He was outraged that when it came to climate change industrial nations placed the burden of proof on small island states like Tuvalu. "[T]he international community would require us to prove to them that climate change is adversely affecting us and what impacts it might have on our economies and what steps can be taken by us before they might consider to act. This is arrogance at its worst. We are not where we are because of our own doing. We are where we are because of the excessive consumption needs and greed of the developed and more powerful countries. They, the developed countries, must accept their responsibility and take the consequences of their actions on others more seriously.

"So it is my plea to the developed countries to take a more responsible and humane stand. You have reached the pinnacle of your might and power and likewise you have all the resources to turn greener and become less greedy for materialism. I plea that you physically take action to reduce now and allow us to continue to live in harmony and in cooperation with nature on our God given lands. You cannot simply continue to shy away from your responsibility whilst we continue to live through the consequences of your actions."

While in the Central Pacific I had a chance to speak with Epel Ilon, the foreign secretary of the Federated States of Micronesia (FSM), a nation comprising some six hundred of the tiny Caroline Islands, the next chain west from the Marshalls. Ilon, a towering, bearded man with a confident smile, has made his share of speeches to conference delegates. The FSM—another former American trust territory—is in only a slightly better situation than its three atoll neighbors, Kiribati, Tuvalu, and the Marshalls. Three of its four "big islands" are younger volcanic cones and the fourth—Chuuk (Truk)—is an archipelago of mountain stubs surrounded by an emerging atoll reef. A 1- to 3-foot sea-rise will cause great hardship to the residents of these "high" islands: Most people live along the narrow coastal fringes that would be inundated by rising seas. But the real tragedy will occur on the hundreds of smaller outer islands and low-lying atolls of the FSM. Entire constellations of these islands could vanish beneath the waves.

"We may not lose our entire nation like the Marshalls or Tuvalu, but in terms of total land area, we have as much or more to lose," Ilon said, sitting back comfortably in his armchair. Through the window at his back, parched grass baked under the blinding sun, waiting for the day when the tiny saplings planted on the grounds would grow tall enough to provide shade. Ilon told me about his home island in the Mortlock Atolls of Chuuk state, how there is sea where, in his childhood, there was land, how temperatures and high tides seem higher than before, and how erosion marches inexorably forward from the shores.

He spoke of flying over the island of Fais during the central Pacific's devastating 1998 drought. Seasonal rains hadn't come that year. While Pohnpei—site of the FSM capital—survived on brackish water, emergency drinking water supplies were rushed to forty outer atolls whose supplies had run out. Fais had been the site of an enormous phosphate mine and had little ground water to start with. As Ilon's plane flew over Fais, the island was obscured by smoke from countless fires. The vegetation had dried and Fais had become a tinderbox.

"We've found that climate change talks are dictated by the economic interests of developed countries," Ilon said. "They worry about what would happen if they were to go back to a 1990 [greenhouse-gas] emissions level, about how that might affect their growth. We think they can and should do much more. Rich countries should be investing in safe practices and technologies to make their industries more efficient and less polluting. International lending institutions could work in partnership with developing countries to foster the creation of clean products and industries. That would be in the interests of all of us, in the interests of the Earth."

The FSM national report on climate contained maps showing sea-rise impacts on the FSM's main islands, rings of settlements and infrastructure surrounding steeply sloped mountains. The villages, roads, airports, and other infrastructure would be largely lost to the sea; the rugged mountains make retreat difficult and, for some facilities, impossible.

"For us this is very urgent," Ilon said calmly. "But there's not much we can do about it, except to talk about it on the international stage. So we'll keep on talking."

∞

ALMOST THE ENTIRE MARSHALL ISLANDS government is housed in the Nitijela, a new post-modern structure in D-U-D that also houses the parliament, for which it is named. There's a public square in front, but it's already choked with weeds, contributing to the premature buckling of the flagstones. As yet, there are no trees to provide shade, something essential to equatorial civics. The parking lot is unpaved and dust hangs in the air long after the infrequent arrival or departure of a vehicle.

Here I met Rhea Moss, the Marshall Islands Undersecretary of Foreign Affairs. She was Marshallese but had been born and raised in Etna, California (population eight hundred), high in the mountains of an expansive continent. She'd graduated from the University of California two years earlier and was only twenty-three years old. While many of her classmates were working summer internships or applying to graduate schools, Moss flew to Majuro to accept a job as a high-level cabinet official in her parents' homeland. University-educated Marshallese are in short supply, and the government values the few who choose to stay or return to the islands.

"I don't think most people here realize climate change issues exist. But they are concerned about the coastline erosion they see and about the inundation of underground water supplies," she said. "They don't know the facts, and even when they do it's hard to think about such things here. I mean this is such a small country and there are *so* many issues that take precedence before sea-level rise."

So she and her foreign ministry colleagues were planning a series of public meetings on the relatively populous Kwajalein and Jaluit atolls, the first step in preparing the nation for its uncertain future relationship with the sea. "For now we just need to let people know that this problem exists. We need to tell them that possibly, years from now, some of the islands will be underwater. That typhoons—

which normally start here and move elsewhere—may stay here and do very great damage to property and lives. We don't really have any tools to arm them with to actually do anything about it. And well, really, what can you do?"

∞

THE SUN BEGAN TO SET OVER THE LAGOON, turning it orange in one bright bloom before vanishing beneath the waves. The arc of islands trailing off towards the glowing horizon slowly disappeared, one by one, into the twilight.

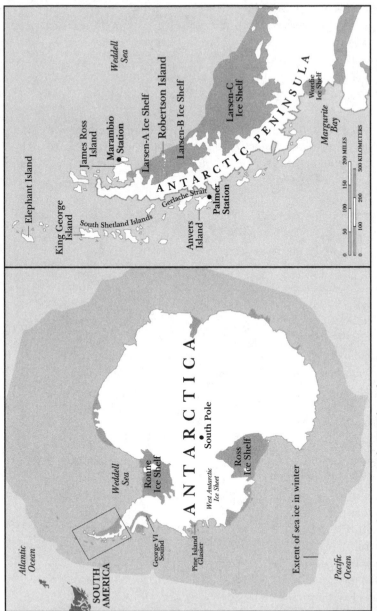

Antarctica & the Antarctic Peninsula

seven

Message from the Ice

On my third morning in Antarctica, Andy Young and I were drilling holes into the melting snow on top of a vast glacier 1,000 feet thick. We used an oversized hand brace, the kind of drill you bear down on with one palm while cranking the handle with your other hand, except this one was nearly 4 feet tall. Every so often Andy climbed up the glacier from the tiny research station below to assess ice conditions, locate new crevasses, and make the appropriate adjustments to the line of flag poles marking the safe trail up to the glacial summit. That was what he was doing today, and I'd offered to help operate the oversized brace to drill new marker holes in the granular glacial crust.

Our thick boots crunched in the wet crust as we marched up the glacier's tongue, a long, steep slope reaching down to exposed rocks behind the station. It would have made a perfect sledding hill if it weren't for those rocks and ledges. A half-mile-long toboggan ride would have ended abruptly amid the primordial landscape of granite that had been scraped, shattered, and sharpened by the enormous weight of the glacier that enveloped the land until a few years ago. The forty-three residents of the U.S. Palmer Station make do with long hikes and cross-country skiing on the glacial plain high above. To get there they have to climb this slope because elsewhere the glacier falls directly to the sea from sheer cliffs of soothing blue ice. Everyone is careful to stay inside the parallel rows of flags Andy sets to mark the safe route.

It was bright and sunny, and after a couple of holes I had to stop and unzip my red government-issue parka. A stream of meltwater gurgled just under the snow pack. Andy was contemplating the pole line, looking like some bearded Heroic Age explorer in tinted sunglasses. The skies were unusually clear that morning, and from our position high on the slope I could see an archipelago of black rocks, ledges, and islands poking out from the frozen white ocean. Here and there, barn-sized icebergs were frozen into the sea ice, some crowned by resting penguins. Huddled together at the end of the rocky point of land jutting out from the base of the glacier, Palmer Station's half-dozen buildings looked preposterously small. Six-hundred-foot cliffs of blue ice loomed over the tiny outpost from the north and east. Further into the interior, sharp, rugged mountain peaks rose out of the vast glaciers, tearing the wispy clouds into whirling white tendrils against the pale blue sky. Here, humans were clearly guests of nature. I wondered how those roles had become reversed over much of the rest of the planet.

Then it happened.

There was a sudden cracking sound—like a giant breaking a redwood over his knee—followed by a rumbling thunder. Garage-sized chunks of blue ice were calving off the face of a glacial cliff directly opposite us across a frozen cove of sea ice—a dramatic vertical avalanche. The blocks tumbled 400 feet into the frozen sea below, throwing up waves that lifted and dropped the 2-foot-thick surface ice with a muted crackling sound as they moved across the harbor. Andy looked up from his work with mild curiosity, smiling at the beauty of the spectacle.

"That was a good one," he said, wiping some ice from his beard. He stuck a tall flag marker into the newly drilled hole. "You can see why we mark the route up the glacier—you just don't want to get too near the edges. Not with the glacier retreating as quickly as it is."

Like many glaciers in this part of the Antarctic, the Marr has been in rapid retreat for the past quarter century. Every year, another 30 to 35 feet of rocky terrain emerges from beneath this mountain of ice. Hills and gullies, boulders and beaches: A stone-carved landscape is quickly unrolling, scroll-like, from beneath twenty thousand years of ice. The prehistoric glacier is melting away like a spring snowdrift.

I took a couple of steps backwards, putting the marker flag between me and the glacier edge. Andy put the oversized hand brace over his shoulder and began climbing toward the next flag.

※

IF IN THE FINAL EQUATION THE SURFACE OF THE EARTH is a single, complex system, then Antarctica is its heart, the slowly beating pump that drives the whole world. Each austral winter, a 7 million-square-mile halo of sea ice forms around the continent, and each spring trillions of tons of fresh water are released into the ocean as it thaws. This is the planet's greatest annual climate cycle, the thermodynamic engine that drives the circulation of ocean currents, redistributing the sun's heat, regulating climate, forcing the upwelling of deep ocean nutrients, setting the tempo of the planet's weather. The Antarctic affects all our lives, but through forces so deep and elemental that we're not even aware of them.

Conversely, here is where global change is most clearly seen. The effects of ozone depletion and global warming are strongest in polar regions, where they are reinforced by atmospheric and meteorological conditions. Because Antarctica is essentially uninhabited and without industry, there is virtually no local pollution; any ecological and climate disturbances on the continent are certainly caused by global forces. The continent is also unowned, and by international treaty it has been set aside for the pursuit of scientific discovery; from outposts spread across the frozen continent, scientists from around the world monitor Antarctica's climate, ice, and animals, assembling a picture of a planet in flux.

For these and other reasons, I packed my warmest clothes and flew south, as far as an airliner could carry me.

※

I BOARDED THE *Laurence M. Gould* in Punta Arenas, a small Chilean port town on the Straits of Magellan, at the very tip of South America. The *Gould* was a brand new ship, built for joint oceanographic research and station supply missions in the Antarctic by the National Science Foundation, the government agency that manages U.S. activities in Antarctica. Antarctic realities had imposed on the ship a peculiar appearance. The ice-reinforced hull was painted bright orange so that she could not be mistaken for an iceberg. The orange bulkheads at her bow stood a full three decks high so as to endure the relentless pounding of Southern Ocean seas; the aft deck was only a few feet above sea level on a calm day, and awash with boarding seas in rough weather.

The *Gould* had proven her seaworthiness on her previous cruise, standing up to days of pounding by 50-foot waves and torrential winds, then breaking through a hundred miles of storm-compressed pack ice to reach Palmer Station. The crossing, which normally takes five days, had taken nine, but the *Gould* had earned the respect of her crew. She needed to. Her architect badly miscalculated her center of buoyancy and, with construction well underway, builders realized the ship would not be sufficiently stable when afloat. To rectify the problem, the yard had to add pontoon-like extensions along the waterline on both sides of the ship. These so-called ice reamers gave the ship an inelegant wake—a second bow and stern wave carved from the sea to keep the ship stable. Rumor had it that this and necessary re-ballasting had decreased fuel efficiency by 40 percent. But the ship had survived a serious storm and would soon weather another.

A variety of heavy lifting cranes and A-frames are installed around the ship, including a 13-ton giant used for lifting fully laden shipping containers on and off docks at Palmer and Punta Arenas. One telescoping crane was mounted inside a special chamber of the ship wherein scientists could mount and fuss with instruments and sampling devices in relative comfort; a 15-foot-tall, hydraulically mounted bulkhead door was then opened and the crane extended to deploy instruments into stormy seas. Inside were three modern laboratories and more computing power than was aboard the starship

Enterprise. Above the bridge was a crown of antennas, dishes, radar, and domes worthy of a naval destroyer. The 230-foot ship was an icebreaker, research ship, and freighter rolled into one.

WE SPENT NEARLY A MONTH AT SEA en route to Palmer Station. Most of my shipmates had come to research the effects of 1998's record ozone hole on the organisms that constitute the base of our planet's food chain: the viruses, bacteria, phytoplankton, copepods, and krill that form the ecological foundation on which all higher life rests.

In all there were twenty-four scientists on board and roughly as many officers, crew, and support staff. There was a mix of ages and expertise: undergraduate chemistry majors, lab technicians with master's degrees, doctoral candidates, post-docs, and five established scientists who had devised the projects that won the grants that brought everyone to the *Gould* in the first place. These senior scientists pored over satellite data to choose our destination. The best place to conduct their research turned out to be the middle of nowhere: a set of coordinates in a violent, unprotected expanse of the Southern Ocean that not only lay directly under the ozone hole but was also renowned for the phenomenal quantity of phytoplankton that grow there. Unfortunately, the area was also infamous for the frequency and ferocity of its storms, which poured out of the Drake Passage every few days, tossing the ship as if it were a child's bath toy. During such a storm, a 6-foot-tall man, asleep on a couch, was thrown completely airborne, colliding with the lounge bulkhead well before his feet touched ground. Forty-foot waves towered over the ship like terrible mountains, then lifted us to their crests so fast that you could feel your blood being pulled out of your head and into your swelling feet. At dinner it was considered polite to catch your companion's meal before it struck the wall or, worse, one of the officers at the next table. It was much like taking an amusement park ride nonstop for four nights and three days.

Our outward trip was calmer. For three days we steamed southwest from South America over gently rolling seas, stopping when we hit a floating snowfield that started with razor edge clarity and continued 500 miles farther south into the Weddell Sea, the seasonal pack ice. There was little transition between ice and sea: The ice edge was a crisp line stretching before the ship as far as the eye could see. Once inside the pack, the temperature dropped 10 degrees F. The entire snowscape undulated with the 3-foot ocean swells accompanied by the shimmering sound of thousands of ice particles stirring around in the slurry between the car-sized chunks. Like the ship, the chunks bobbed slowly in the swells, sinking a couple of feet, then rising as if on tip-toe to reveal ice blue undergarments. Snow petrels— whiter than white—circled the ship in tight circles like pigeons. Dozens of crabeater seals slumbered on the ice pack as if close encounters with huge orange ships were an everyday occurrence. This first glimpse of the Antarctic was at once beautiful and dangerous. The swells bounced boulders of ice against the hull, which shuddered from the force. We turned back to open water.

For two weeks we remained in the open Weddell Sea, exposed to the fury of the Southern Ocean, as scientists collected water samples and conducted experiments. Some days were so rough that all science was stopped and the captain ordered that nobody was to go out on deck for any reason. On other days, experiments and instrument deployments continued around the clock. Every day, a misty, overcast sky hung above the angry gray sea. Cape petrels and albatross remained constant companions, circling the ship on the cold polar winds, perhaps in the hopes that we would eventually turn out to be a trawler. I read *The Rime of the Ancient Mariner* and resolved to be kind to albatross throughout the journey. Never did we come within sight of land.

Small icebergs passed us regularly. Watching them became a constant worry for the crew and a popular pastime for the rest of us. They came in all shapes and sizes—from "bergy bits" no bigger than a piece of household furniture to tabular islands several miles across. The most dangerous were heavy boulders of older ice, called "growlers" because of the sound they make when their tops bob out of the water. These weigh several tons but are difficult to spot amid whitecaps and towering seas. The bridge crew was engaged in a constant slow-motion slalom exercise to avoid the wind-driven bergs bearing down on the ship. I asked the captain—a respected Louisiana Cajun named Warren Sanamo—if the ship's reinforced hull would protect us. "Oh," he smiled, "ice like that could put a fair dent in a hull like this."

Some of the larger bergs may have been remnants of the Larsen-B and Wordie Ice Shelves, ancient platforms of floating ice hundreds of yards thick that collapsed earlier in the decade. Freed from their moorings, they had drifted slowly out into the Southern Ocean and were gradually melting away.

☞

THE ANTARCTIC PENINSULA REGION is a thousand-mile arm of glaciated land and islands, reaching northward towards the tip of South America, and it is undergoing a dramatic realignment of its climate. Over the past fifty years average temperatures in the region have increased by almost 5 degrees F. This is a shift of staggering proportions. By comparison, average global temperatures are only 7 to 11 degrees F higher today than at the height of the last great ice age, when Britain, Scandinavia, and much of North America were blanketed in mile-thick ice.

Thus far, two large ice shelves have collapsed on the Antarctic Peninsula as a result of this warming. In the 1980s, the Wordie, the northernmost ice shelf on the western side of the peninsula, completely disintegrated. On the opposite side of the peninsula, large portions of the northernmost ice shelf—the Larsen-A—began breaking off into the Weddell Sea. Soon, researchers detected increased quantities of icebergs in the Southern Ocean. Then, in early 1995, the Larsen-A collapsed with a suddenness that shocked the scientists who monitor these events.

Rodolfo del Valle, director of geoscience at the Argentine Antarctic Institute in Buenos Aires, knew the Larsen-A Ice Shelf like the back of his hand. He had traveled up and down it for twenty years, studying geological formations along the eastern coast of the peninsula. Larsen-A was an Antarctic geologists' dream: a relatively flat, thousand-foot-thick platform of ice and snow, roughly the size of Delaware, bordering a rugged, mountainous, largely unstudied landscape.

"Larsen-A was our highway," del Valle recalled over a crackly phone line a couple of weeks before I left for Antarctica. "It allowed us to use [snowmobiles] to travel between a great many points of geological interest." His team set up small fuel and supply depots all along the massive shelf, which began as a thin tongue connecting James Ross Island with the mainland and slowly widened over its hundred-mile length. Towards the southern end—where a chain of islands separated it from the larger Larsen-B—it was nearly 60 miles wide. Fed by mountain glaciers, the Larsen-A had covered this stretch of the Weddell Sea for many thousands of years. Only a handful of tiny islands poked their rocky peaks above the ancient blanket of floating ice; these so-called nunataks provided convenient bits of dry land where the geologists set up their seasonal field camps and stored supplies during the long Antarctic winter.

The stability of the ice shelf had been a concern for a number of years. A portion of its ice front had been in constant retreat from an area north of Robertson Island since the 1940s, and the thin tongue linking James Ross Island with the mainland became separated from the rest of Larsen-A in 1958. The retreat accelerated after 1986 as mean summer temperatures at Argentina's nearby Marambio Station hit record highs; during the 1992–1993 austral summer, average air temperatures were actually above freezing. The edges of the ice shelf continued to retreat. Deep cracks and crevices appeared on its surface. The summer of 1994–1995 broke all previous temperature records at Marambio, with an average of +0.6 degree C (1.1 degree F). Scientists worried that, like the Wordie a decade before, the entire shelf might disintegrate in a few short years.

The collapse happened in a few short *days*.

Del Valle and his colleagues were on a nunatak one stormy day in January 1995.

"We were expecting a break-up, but not so fast," del Valle said. "There was an enormous sound like thunder all around us. There were huge waves and everything trembled and shook. We watched in shock as the shelf collapsed around us."

Two hours later, the thirteen Argentineans found themselves standing, not on a rocky outcropping, but on an island surrounded by open water and enormous icebergs.

"I felt a sadness, a pain in my heart, for the loss of a place that had become like a home to me. I've experienced strong earthquakes on land, but this was different. After an earthquake something remains. But not with the ice shelf—it was completely destroyed. I felt like I'd lost my own country." Surveying the damage by helicopter, the first thing del Valle did was cry.

The Argentineans had witnessed one act of the dramatic collapse of the ice shelf, which occurred over several days. To the south, an iceberg 48 miles long and 23 miles wide calved from the face of the Larsen-B. The orphaned ice shelf connecting James Ross Island also disintegrated, opening Prince Gustav Channel for the first time in recorded history. (Two years later, del Valle was aboard the first ship to circumnavigate the newly freed island, a Greenpeace icebreaker publicizing the impacts of global warming.) In all, 1,620 square miles of ice 1,000 feet thick had collapsed into the sea.

∞

AT THE END OF OUR FIRST WEEK in the Weddell Sea, flurries of concerned e-mails arrived from friends and relatives. A few hundred miles further south across the Weddell's pack ice, a tabular iceberg the size of Delaware had calved from the Ronne Ice Shelf. At that latitude it is sufficiently cold to allow permanent platforms of ice to spread from glaciers to cover great stretches of the sea, year-round, to a thickness of hundreds of feet. The Ronne Ice Shelf, a thousand feet thick and covering 180,000 square miles of the inner Weddell Sea, is generally thought to be tens of thousands of years old. It is a floating ice shelf, like the Wordie and Larsen, but a hundred times larger.

Once detected by an American satellite, this enormous calving made headlines around the world. German scientists soon discovered that one of their research outposts, the then unmanned Filchner Station, was adrift on the new iceberg and would have to be salvaged by ship. The *Gould* was the vessel closest to the event, but it might have taken several years for the berg to reach our position.

At this writing, attention has turned to the much larger Larsen-B, now the northernmost ice shelf in all of Antarctica. It's about the size of Connecticut, or at least it was. In early 1997, cracks the size of small canyons were detected in the shelf. In March 1998 an iceberg 24 miles long and 3 miles wide broke away from the shelf, and scientists began predicting its imminent collapse. It would be as large as all previous ice shelf collapses combined. The shelves appear to be vanishing sequentially from north to south.

As dramatic as they are, the losses of these ice shelves won't influence world sea-levels—at least not directly. Because they were already floating, they've been displacing their entire mass in the ocean for millennia. But they may be a harbinger of much more unpleasant events ahead. If the worst possibilities came to pass, the Marshallese would not be the only people affected.

༄

ANTARCTICA, OF COURSE, IS COVERED IN ICE. The continent is twice the size of Europe, and 98 percent of it is buried in glaciers and other year-round ice cover. The glaciers I visited on the Antarctic Peninsula are mere dwarves when compared with the main ice cap, which covers the continental interior to a depth of 1 to 3 miles. It is so heavy that it has depressed the bedrock underneath, sinking the land by as much as a mile so that large portions of the continent are actually below sea level. Ninety percent of all the fresh water on Earth is locked up here. If it ever melted entirely, the oceans would rise by 200 feet.

Scientists have predicted since the late 1970s that climate change impacts would be most extreme in polar regions. In a famous 1978 paper in the journal *Nature*, J. H. Mercer suggested that Antarctica would serve as humanity's early warning system and cited the collapse of the peninsular ice shelves as among the first indicators of global warming.

Around the same time, scientists first started worrying about the stability of the West Antarctic Ice Sheet.

To find the Ronne Ice Shelf, follow the Antarctic Peninsula south from the open waters that once were covered by the Muller, Wordie, and Larsen-A Ice Shelves, past the collapsing Larsen-B, and you can't miss it. The Ronne, roughly the size of Spain, covers the entire head of the Weddell Sea. As an ice shelf, it is by definition floating, but it towers over the surrounding sea ice like the white cliffs of Dover. If the main body of the Larsen collapses, the Ronne will be the northernmost ice shelf on the eastern side of the peninsula.

Ice, like water, falls towards the sea, but it does so extremely slowly. The Ronne Ice Shelf is fed not by a glacier but by the continuous, slow-moving spillover from the western component of the south polar ice cap: the West Antarctic Ice Sheet, or WAIS, as it is generally called. WAIS's internal ice streams typically flow towards the sea at a rate of a few hundred yards per year, spilling out over the Weddell Sea. On the other side of West Antarctica, WAIS is also spilling into the Ross Sea, where it forms the Ross Ice Shelf, the Ronne's equally large cousin to the west.

WAIS is an enormous slab of ice many, many times older and more massive than any of the floating ice shelves. Indeed, both of the world's largest ice shelves are mere byproducts of this miles-thick ice cap, which covers 760,000 square miles, an area the size of Mexico. This vast quantity of stored water is precariously balanced. Its weight has depressed the underlying bedrock sufficiently that its bottom lies below sea level. This makes it the world's last surviving marine ice sheet, its companions in the Arctic having collapsed tens of thousands of years ago.

Because ice floats, marine ice sheets are inherently unstable. They survive only as long as they have sufficient mass to prevent sea water from intruding under the base. If regional warming were to bring about the disintegration of the Ronne or Ross Ice Shelves, scientists warned in the 1970s, it might destabilize WAIS itself. Sea water working its way under the bottom of the ice sheet could trigger a catastrophic break-up. The weight of the ice sheet, currently borne by bedrock, instead would be displaced by the sea, into which it would eventually melt.

In this scenario, world sea levels would rise by about 18 *feet* in as little as a century. That's in addition to the 1- to 3-foot estimate for the twenty-first century provided by the IPCC. Most coastal regions and hundreds of cities, including New York, London, and Jakarta, would be submerged. Such a disaster would place enormous strains on world civilization from the loss of cities and farmland, changes in currents, and the destabilization of weather systems. It's the "apocalypse scenario" for global warming.

Over the past twenty years the U.S. and Great Britain have financed extensive research into the stability and mass balance of WAIS. Glaciologists and geophysicists have probed, modeled, drilled, and measured the ice sheet in an effort to assess its stability and the rate at which it is shrinking. To study the ice sheet's myriad relationships with its component ice shelves, with historic and recent climate fluctuations, and with rates of iceberg calving in the Weddell and Ross Seas, field camps have been erected on top of giant ice streams that pour from within WAIS toward the ocean. Today we have a far clearer picture of how the ice sheet functions.

The good news is that scientists are fairly certain that WAIS won't collapse during the next century. There are several reasons for this optimism. First, climate models suggest that the Ronne and Ross Ice Shelves would not disintegrate unless mean air temperatures in the area increased by at least 5 degrees C (9 degrees F). This is unlikely to happen anytime soon since it represents a warming trend twice as severe as that now being experienced along the Antarctic Peninsula, hundreds of miles north of these larger ice shelves.

Second, there is preliminary evidence that WAIS's stability may not actually depend on the presence of these floating ice shelves after all. The glaciers that fed the Wordie Ice Shelf have shown little sign of accelerating retreat since the shelf collapsed. Glaciologists are also looking at a weak point within the WAIS system itself, a retreating coastal glacier called Pine Island glacier. During the last ice age, the Pine Island glacier is thought to have fed a floating ice shelf that has since disintegrated. In the absence of this ice shelf, the Pine Island glacier is retreating 0.7 mile per year. This might create exactly the sort of weak spot through which WAIS might begin collapsing into the sea. To everyone's relief, thus far none of WAIS's seaward flowing ice streams appear to have increased their flow in this direction.

Charles Bentley, a geophysicist at the University of Wisconsin who has worked on WAIS since the 1950s, says that it is not at all clear what would happen if the Ronne or Ross ice shelves thinned and disappeared. "Pine Island Glacier is rapidly retreating, and yet there's no indication of a disaster in the making," he points out. "This tends to support the idea that the ice shelves aren't so crucial to the health of the interior ice sheets."

Similarly, in studying the ice streams within WAIS, Bentley has found that there are a great many *internal* shifts and changes within the ice sheet: Some ice streams shut down, others speed up, some change their boundaries and routes. But taken as a whole they appear to cancel each other out. "There doesn't seem to be a major increase in overall output," he says. In the short term, he argues, the chance of a catastrophic collapse is exceedingly slim: 0.1 percent in the next century or two.

That's not to say that WAIS is stable: Geologically speaking, it indeed appears to be a system in the process of collapse. What Bentley and other researchers argue is that the collapse will probably take a few thousand years.

Still, researchers really don't know how quickly WAIS is breaking up, nor do they know how the ice sheet will respond to a continued warming of the global climate.

"We don't think there will be a catastrophic collapse in the next century, but what about a 5 or 10 percent loss?" David Vaughan, a glaciologist with the British Antarctic Survey, said to me before I left for the Antarctic. "A loss of 10 percent of the ice in WAIS would translate to another 50 centimeters of global sea level rise."

That would nearly double "best guess" sea-level rise projections for the next century from 50 centimeters (20 inches) to almost a meter (3.3 feet). The IPCC currently assumes that there will be virtually no net contribution from Antarctica: increased snowfall will balance any losses from WAIS. "But they state in their full report that this estimate is not based in known scientific principles," Vaughan notes. "It's based on a lack of knowledge. They just put in a zero."

Charles Bentley more or less agrees. "Let's suppose the WAIS were to retreat at a steady rate so that in five thousand years it would be gone. That means five meters of sea-level rise in five thousand years or one millimeter a year. That would correspond to a 50 percent increase in the current rate of sea-level rise. And I think that's about as fast as I'd expect the ice sheet to disappear. Whether or not you'd consider that instability depends on whether you're speaking in human or geological time scales. So while I don't think we need to worry about the WAIS collapsing, that's not to say what's happening is totally unimportant. It could be significant for sea-level rise in the future—maybe so much as doubling it. That's significant—maybe not catastrophic, but significant."

There's no way of knowing if WAIS will shed its mass at an even rate. "One millimeter per year is just an average figure," pointed out another glaciologist, Robert Bindschadler of NASA's Goddard Space Flight Center. "There's evidence that there hasn't been just a steady retreat, but that there have been times when the sheet retreated quickly and other times when it moved more slowly. It looks like we're in a period where it's retreating more slowly—I'd guess it's not more than half a millimeter per year addition to world sea levels."

It's still too early for any of the WAIS scientists to say what causes the enormous ice cap to quicken or slow its bleeding—or if the disintegration of the adjacent ice shelves would have an adverse effect on WAIS's stability. For everyone's sake, one hopes this icy behemoth dies a quiet, natural, and very, very slow death.

But while glaciologists ponder WAIS's future, Antarctic biologists are recording dramatic climatic impacts of a different sort. The warming of the Antarctic Peninsula is having a clear and extremely troubling impact on life around its shores.

ᘓᕈ

ONE MORNING AFTER SEVERAL WEEKS in the open ocean, I stepped out on deck and was jolted by the sight of mountains looming over the ship. During the night, the *Gould* had taken up station in the lee of Elephant Island to avoid an approaching storm. The island blotted out the northern horizon with peaks so high and jagged they might have been created the day before. Much of coastal Antarctica looks as if somebody flooded the Rockies or Swiss Alps up to their glaciated peaks. Two-thousand-foot drifts nearly bury the craggy mountains and end in sheer drop-offs where they calve into the sea amid grounded icebergs. Elephant is an uninhabited and remote island standing sentinel 125 miles north of the Antarctic Peninsula, an ominous foreshadowing of a rugged, hostile continent beyond.

It was here that twenty-two members of Ernest Shackleton's *Endurance* expedition survived for 105 terrifying days in 1915. After several months trapped in the Weddell Sea ice pack, their ship had been crushed. They'd already spent 170 days drifting on unstable ice floes without adequate food or shelter, followed by 7 frightening days in the Southern Ocean in the overladen open lifeboats they'd dragged across the ice. "Such a wild and inhospitable coast I have never beheld," the expedition photographer wrote in his diary after their precarious landing on Elephant Island. With no hope of rescue and winter setting in, Shackleton and five men set forth on a suicide mission: to cross 800 miles of open ocean in a 22-foot open lifeboat to the whaling station on South Georgia Island. Their 16-day crossing and subsequent 36-hour climb over uncharted, 3,000-foot mountains to the whaling station is Antarctica's greatest story of human endurance. In the end, all twenty-six members of the expedition survived their odyssey.

Once the *Endurance* had been crushed in the floes, Shackleton's party became dependent on whatever food the Antarctic could provide. Vegetarians would have done poorly. For nearly two years the castaways survived almost entirely on seal and penguin meat. In the final months, the men marooned on Elephant Island were reduced to an all-meat diet and were weary from the perpetual slaughter. "About thirty Gentoo Penguins came ashore and I am pleased the weather was too bad to slay them," one member scrawled in his diary towards the end. "We are heartily sick of being compelled to kill every bird that comes ashore for food."

Penguins are not only flightless and awkward on land, evolution also has conditioned them not to fear terrestrial creatures. This makes them remarkably easy to catch. Untold millions of penguins were killed for food and oil by whalers, sealers, and Antarctic expeditions. In the northern subarctic, hunters and fishermen completely exterminated the great auk, a distant relative of the Antarctic penguins. There are now no penguins at all in the Arctic and subarctic.

Aboard the *Gould* we were liberated from such slaughter by walk-in freezers, but penguins provided considerable solace from the monotony of shipboard routine. I had my first close encounter with wild penguins on a raw, overcast day. We were taking water samples in rolling seas some 80 miles north of Elephant Island when two Chinstrap penguins appeared beside the ship. Man and penguin met face-to-face. I brought to this encounter a lifetime of media and advertising images of "cute" penguins that had engendered deep-seated cynicism, a resolve not to go cooing over penguins, the Eiffel Tower, Waikiki Beach, or any other pre-packaged pop icon.

To my horror, I found them completely, inexplicably charming.

The Chinstrap penguin is a foot-and-a-half long with a white chest and black body and cap. It has a distinctive black line under its chins that appears to some to be holding a black cap onto its head; I thought it looked like they were wearing sleek, white racing masks on their faces. They popped up out of the rollers alongside the deck where I stood, expressing a great deal of enthusiasm for the stationary ship. One would pop up, shake its head as if in disbelief, first at me, then at its newly surfaced companion. Then they'd both dive underwater and take off, wings flapping rapidly and gracefully beneath the water, ducking under the ship, turning pirouettes as they orbited the vessel at incredible speeds. They moved like undersea barn swallows. Then—pop!—onto the surface again, ruffling with enthusiasm. They continued this behavior round the clock for three days.

At first, it seemed they were feeding on something the ship was stirring up with its propellers—although what that would be in 7,000 feet of water was beyond our party's considerable oceanographic knowledge. But that afternoon they became fixated on the ship itself. Geoff MacIntyre, a technician from Prince Edward Island whose love for the outdoors had him on deck in all conditions, was standing on the stern in a survival suit. He was reeling in an instrument by hand with the stern doors open when the Chinstraps popped up right under his feet, swimming around like excited pets greeting their owner on his return from work. "I thought they were going to jump me," he said afterwards.

The next day I caught them trying to board the ship. They'd get themselves going at 20 miles an hour under the water, then soar out of an approaching wave and onto the *Gould*'s ridiculous "ice reamers." Unfortunately, the tops of the reamers are angled downwards, and the penguins would promptly slide back off them into the sea with a plopping sound. They showed no sign of distress, but continued their efforts to alight on our noisy, undoubtedly alien-smelling vessel throughout the day. The next morning, Randy Sliester, a wilderness enthusiast from Colorado, was deploying instruments from the Baltic Room, the *Gould*'s special crane-equipped chamber. Randy was in charge of the ship's scientific support staff and was on hand for most instrument deployments. A deep-ocean instrument assembly was several thousand feet down collecting water samples and Randy was waiting patiently in the open two-story tall Baltic Room door, tied to the ship with lifelines and watching the surf break just inches below the lip of the bulkhead doorway. "Then there they were, all excited and stuff. They started making for the door, trying to jump into the Baltic Room," he said. "They didn't make it, but they ignored me when I tried to discourage them. I was on another cruise down here and a penguin actually shot out of the water and into the Baltic Room. He started running around all freaked out and stuff, flapping his wings and squawking. And we're all stumbling around in our mustang suits and lifelines trying to catch him or herd him back out the door."

I sent an e-mail to Bill Fraser, a penguin researcher from Montana State University who has worked at Palmer Station for nearly three decades. Neither he nor his colleagues at Palmer had ever heard of such behavior. "They may just be curious," he suggested.

In the Gerlache Straits we watched Gentoo penguins porpoising through the water like awkward dolphins. Adélies would watch the ship with indifference, then take off in their Charlie Chaplin waddle, flipper-wings flapping behind them. (In Chile, I visited a Magellanic penguin rookery and had three birds cautiously approach me as I sat on a popular commuting path to the beach; all three walked right over my boots, eyeing me with a mix of curiosity and suspicion, and then ambled on, arms outstretched for balance.)

By the end of my trip, I couldn't look at a penguin without smiling or, more often than not, breaking into laughter. Maybe they have this effect on people since their movements resemble those of toddlers. When they stand still they evoke an image of a poised butler; on land they're the most anthropogenic of animals. In their underwater flight they are as graceful as terns.

They are also adapted to survive in one of the most hostile environments on Earth. Emperor penguins, the largest and hardiest of the lot, can actually lay their eggs on bare ice; the males then incubate them in total winter darkness, unable to eat for 115 days while standing in –50 degree F temperatures and being battered by 180-mile-an-hour winds. At the end of their fast, they turn the hatchling over to the returning female and must waddle as far as 60 miles to open water to feed.

All these penguins are descendants of flighted birds that remained on Antarctica tens of millions of years ago as the continent drifted southwards. To survive in an increasingly frigid, lifeless land the birds adapted to fly not in air but in water. Light, hollow bones became solid and heavy. Their wings became rigid flippers, their feathers transformed into a furry down. As terrestrial food supplies became scarce, penguins captured a vital ecological niche. While other seabirds are confined to pursuing food at the ocean's surface, penguins regularly dive 300 feet into the depths in search of food: krill, in the case of most species, although some hunt squid or fish.

Some range 70 miles from home each day in search of food—a long swim for any creature, much less a flightless bird.

Bill Fraser first came to Palmer Station to study penguins in 1974. It was a perfect location for this sort of research. Chinstrap, Gentoo, and especially Adélie penguins nested in large, bickering colonies on many of the rocky little islands dotting the icy seas around the station. Slowly, he began learning how penguins, skuas, petrels, and kelp gulls live and die. He has been doing so every summer since.

"My project began at Palmer Station as a very basic undertaking to understand the biology of sea birds," he told me from his office in penguinless Bozeman, Montana. (Our respective visits to Antarctica that season did not overlap.) "Through a lot of luck we ended up securing funding for a very long period of time, which let us develop very long-term databases on the demography—the population dynamics—of most of the species there, particularly penguins. We had no way of knowing that those databases would become indicators of climate warming on the Peninsula."

Because sea ice figures so largely in their lives, penguins are excellent barometers of climate change. Adélies, Chinstraps, and Gentoos are closely related, similar in size and appearance. All three feed on swarms of Antarctic krill, the inch-long shrimp-like creature that forms the basis of the Southern Ocean food chain. Unlike Emperors, they all lay their eggs in spring on nests of pebbles gathered on dry rock. The males fast until relieved at the nest by foraging females, who often travel great distances to deliver regurgitated food to the chick, which hatches just prior to her return. If the female returns late or not at all, the chick is doomed. The male will abandon the nest and attempt to find open, krill-rich water before starving to death.

Timing is everything. The snow must thaw from nesting sites by the time the females lay their eggs; the sea ice edge must be within the foraging range of the females; the krill must be swarming on her arrival; the chick must hatch and be fed before the fasting male is forced to abandon the nest.

While similar in these respects, the three peninsular penguin species have some crucial differences. Chinstraps and Gentoos, as I saw on the

Gould, are maritime creatures, capable of feeding in open water for days on end. They prefer open water and time their breeding later than Adélies to ensure that the ice pack will have broken up during the females' foraging run. In winter, they migrate to the expanding Weddell ice pack where they stick to open water. Gentoos, for whom the peninsula is at the southern end of their range, dive deeper than the others and will hunt fish as well as krill in the open sea.

By contrast, Adélies are penguins of the deep south. As such they are adapted to life on the floes, as they must make long hikes across the pack to get around in the more hostile climates farther south. They depend on the presence of sea ice during winter, as they must hop out onto land each day to rest. During winter, while Gentoos migrate north and Chinstraps stick to open water, Adélies tend to congregate in dense pack ice, 10 to 20 miles from open water. They prefer polyanas, areas of open water within the ice pack in which they can feed on krill by day and find ample resting ground at night.

The ice-dependent Adélies are increasingly finding themselves out of luck.

Bill Fraser is amazed by the changes he has witnessed on the Antarctic Peninsula. In twenty-five years he has seen glaciers retreat to reveal new islands and peninsulas, vast tracts of ice-free land, and dramatic changes to once-familiar vistas and landscapes. "When I was a student I remember that climate change was envisioned as a slow process that could not be observed in a human lifetime. But that's very wrong," he said. "Climate change can be a very rapid event, with abrupt and dramatic consequences for the landscape and ecology."

A warmer peninsular climate means less ice, and not only in glaciers and ice shelves. It also affects the seasonal ice that forms like a gigantic halo around the Antarctic every austral winter. In April and May, this halo spreads at a rate of 25 square miles per minute, ultimately covering an area twice the size of the United States. Or at least it did.

In the area around the Antarctic Peninsula, winter sea ice appears to be on the decline. It turns out that the peninsula's 5-degree F annual mean temperature increase is not spread evenly throughout the year. While summertime temperatures have changed little in the past fifty years, mean mid-winter temperatures are up 9 degrees F. A few decades ago, relatively cold winters brought extensive sea ice—defined as reaching as far as 60° 50' S—in four of every five winters. Now only one or two winters of every five are cold enough. Recent studies of catch records left by whaling ships (the vast majority of whose kills were made within a few miles of the sea ice edge) indicate that the average extent of annual sea ice shrank by 25 percent between the mid-1950s and 1970. Reliable satellite data available since the early 1970s indicate that sea ice in the western Antarctic Peninsula region declined by a further 20 percent through the early 1990s.

"Winter warming has dramatically affected the availability of penguin habitats," Fraser said. "The Adélies' winter habitat—sea ice—has decreased. The Chinstraps avoid ice and their habitat has increased. It's caused a very marked change in their relative numbers." Since surveying of the Palmer area began in 1975, Adélie populations have fallen by 40 percent, from 15,200 breeding pairs to 9,200. On the small islands around the station, twenty-one colonies appear to have gone extinct. Fraser's Montana State colleague Wayne Trivelpiece has recorded similar Adélie declines at his long-term field camp on King George Island, 200 miles to the north.

Over the same period the ice-avoiding Chinstraps have seen a six-fold population increase at these two sampling areas, although they've recently stabilized. Gentoo survey data don't go back as far, but Fraser says these open-water penguins appear to be in the early stages of a population explosion of their own.

Similar shifts are occurring among the region's seals. Weddell and crabeater seals rely on pack ice and are declining in numbers. Populations of elephant and southern fur seals—both open-water species—are dramatically increasing and extending the southern part of their range. When Fraser began surveying Palmer wildlife in 1975 he found only six fur seals in the area; in 1995 there were two thousand. Elephant seal numbers had tripled.

"The story is really about one ecosystem replacing another," Fraser concluded. "In scaling that down a little you can envision one seabird community replacing another. If these trends continue then at some point in the future there will be a very different seabird community in the area than there is currently. It's hard to predict where things will end up, or whether it's a good thing or a bad thing. But I can tell you that there are very significant changes occurring in this ecosystem."

∞

LIKE MANY OF ANTARCTICA'S LIVING CREATURES, Langdon Quetin's life revolves around krill. Each austral summer, he travels halfway around the world to observe and capture these little crustaceans upon which so much marine life depends. He nets them, dives in frigid waters to film them, probes their secrets in laboratories at Palmer Station and the University of California at Santa Barbara. He is married to another of the world's top krill scientists, Robin Ross. During the summer research season they usually come to Antarctica in separate shifts, alternating care of their only child. In the long-term pursuit of krill, they've adopted a penguin-like lifestyle.

It wasn't until our fourth and final week at sea before reaching Palmer that I finally had the chance to speak with Langdon about krill. He'd been hibernating for much of the cruise, on which he was a reluctant passenger; he would have to spend four weeks with the *Gould* to spend four weeks doing his work at Palmer. It was, unfortunately, the only way to get there. He slept during the lengthening spring days, emerging only at night. When the computer lab quieted down for the evening he settled in front of a computer to carefully analyze rare video of krill swarming beneath the ice floes to feed on the algae that grows on and within the ice itself. Only when we turned south from Elephant Island and began making our way down the peninsula toward his research site did I find him up and about before lunchtime.

We'd been talking in the rolling mess hall for only a few minutes when he looked at me with an expression of utter amazement. "You mean you've never *seen* one?" he said. "Well, come on!" He jumped out of his seat and marched me straight out to the 01 Deck where large water-filled Plexiglas incubators had been set up to run in situ experiments on how ultraviolet light exposure affects microscopic organisms. We were, at this time, under the largest ozone hole ever recorded over the Antarctic.

But instead of submerged sample vials filled with unseen microbes, Langdon presented a recent addition to what the scientists perversely termed "the incubators." There, in oversized test tubes, were a half dozen live krill.

So this was the cornerstone of life in the Antarctic, the creature upon which all higher life depends, the short link between microscopic diatoms and the largest animal on Earth. Penguins, seals, baleen whales, and the few remaining blue whales are all sustained by vast swarms of these tiny crustaceans. Inch-long, nearly pinkish, miniature shrimp, so thin that their stomach contents could be clearly seen through their shells. "It's the dark section right behind the eyes," Langdon stuck his finger into the icy water to point it out. "If it's clear, they're eating copepods. Different kinds of phytoplankton turn it different shades of greens or browns. If they're eating other krill it's red." Nature's palette, distributed in sample-sized living tubes.

What they lack in size krill more than make up for in quantity and fecundity. There are an enormous number of krill on Earth—600 million million of them. They are thought to have a combined mass of 650 million tons, far more than the combined weight of *Homo sapiens*. It's a good thing, too. The once-numerous blue whale—a creature weighing 200 tons—eats 6 tons or 6 million individual krill each day. Seals and seabirds, penguins and whales all feed on krill swarms, which sometimes grow so large they turn the sea red for miles.

Krill are fairly long-lived: They can survive for up to seven years, although the odds of being eaten first are quite high. They lay their eggs in huge quantities over the abyssal plains of the Southern Ocean. The eggs sink to great depth and hatch into larvae, which rise from the depths and live hidden under the sea ice, where they find protection and food. Several varieties of algae grow on and within the underside of polar ice, eking out a photosynthetic existence in the dim twilight. The sea ice is a gigantic krill nursery, the Southern Ocean equivalent of a Louisiana marsh or a tropical mangrove swamp. Less sea ice means less nursery habitat and, presumably, fewer krill.

"We actually think krill are dependent on sea ice in two ways," Langdon was saying. "One is that the larvae need the sea ice to feed during the winter—they're obligatory feeders of under-ice algae. There's just not enough other food for them in the water column. We also think that in spring the adult females must feed on the blooms of algae released from the melting ice in order to become reproductive." So less ice also means fewer eggs. "Females aren't ready to reproduce until their second year, so the best situation for krill abundance is to have two years in a row of good, extensive pack ice. The first year would ensure that a large number of larvae survive to adulthood. In the second year, the females would probably come into reproductive condition earlier and would perhaps cycle more eggs so you might have an even bigger year-class."

Now that there are fewer years with extensive sea ice, there's some evidence that krill numbers are shrinking. In a 1997 *Nature* article, Valerie Loeb of the Moss Marine Landings Laboratories and six other scientists correlated reduced sea ice with a ten-fold decrease in average krill populations in the period 1984–1996. Such a decrease in food supply would likely contribute to an overall decline in penguin numbers.

But Langdon's not so sure that krill populations are really crashing. "There's good data only for about ten years. There was a trend going down, but the recent trend is going up. We can correlate krill numbers to sea ice conditions, but if the natural ice cycle is five to eight years and krill can live to seven, it's going to take a couple of decades before we can get a good feel for it." The ten-fold crash could be part of a natural cycle. Factors may be at play other than sea ice extent: when and where the ice forms, how quickly, what shapes it winds up in. So many questions, so many research lifetimes required to answer them.

While we were talking, one of the krill in the incubator stopped moving. The others were cycling their little rows of feet much more slowly. When I came back on deck later in the day they'd turned a deeper pink, like cooked shrimp. It was a sunny day, but the water in the sampling tank had been kept at a sub-freezing Southern Ocean norm of 30 degrees F through the constant pumping of ocean water through the tanks. The sun had killed the krill, not through infrared heating but by ultraviolet burning.

∞

THE SCIENTISTS ABOARD THE *Gould* constantly, anxiously tracked the path of the spring ozone hole far overhead in the stratosphere. Images were downloaded from satellites. Spectrometers on the crow's nest tested sunlight for increased levels of higher-energy UV light, the so-called UV-B that a properly functioning ozone layer filters out. This high-energy light is capable of damaging all sorts of living tissues; it is exposure to UV-B radiation that causes severe sunburns. Most creatures lack proper protection from UV-B since ozone molecules in the upper atmosphere have shielded life below. But here in Antarctica the springtime shield is broken, and scientists are wondering what that's doing to the tiny creatures that form the bottom of the marine food chain.

The now infamous ozone hole was created by mankind's production and release of chlorofluorocarbons (CFCs), a family of chemicals used in air conditioners, refrigerators, and aerosol cans. When they were invented in the 1920s they seemed the perfect gases: They wouldn't burn, explode, corrode, or poison. For fifty years, nobody realized that once released into the atmosphere, CFCs are perfect ozone-destroyers. Once in the stratosphere, UV radiation breaks CFCs apart, freeing chlorine atoms. Chlorine, in turn, steals oxygen atoms from ozone, turning it into oxygen gas (O_2), which does not protect the Earth's surface against UV-B. Each chlorine atom destroys an average of one hundred thousand ozone molecules before itself breaking down or settling out of the atmosphere. This reaction works best in cold conditions, which is why the first and oldest hole began appearing over the coldest part of the Earth, the Antarctic. More recently, scientists have detected a thinning of the ozone layer over the North Pole; some think it could develop into a seasonal hole similar to that over the Antarctic.

Fortunately, production of CFCs is now restricted under an international agreement, the 1979 Montreal Protocol. But the degradation of the ozone layer is expected to continue for many years as the considerable stock of CFCs in the lower atmosphere slowly migrates up to the stratosphere. The problem will get worse before it gets better.

A couple of weeks before I left for Antarctica, satellites above the frozen continent recorded the largest ozone hole in history. On September 19, 1998, the hole in the ozone layer had expanded to 10.5 million square miles. It extended over the entire continent, and as the Earth rotated, wobbling arms reached out over the end of South America. Every few days, Punta Arenas and Ushuia, Argentina, were bathed in unprecedented levels of UV-B as the hole passed overhead. Antarctica experienced these effects almost continuously for two months.

While there's little life in the Antarctic interior to be damaged by this springtime bath of harmful radiation, the penguins, seals, krill, copepods, and phytoplankton that live along the Southern Ocean shores live in harm's way. Increased UV-B has been shown to kill or stunt starfish larvae, urchin embryos, even adult krill. The often exposed backs of humpbacks and other whales that migrate to the Southern Ocean are more regularly seen to peel from acute sunburn.

The scientists aboard the *Gould* were concerned about the ocean's smallest residents: the bacteria, phytoplankton, and microscopic animals that form the foundations of all marine ecology. Phytoplankton and other algae are the food of krill. If they are being harmed by UV-B, there might be less food for krill. There might be less life to go around. Or it might not make much difference. That's what they're trying to work out.

The chief scientist on this particular cruise was Patrick Neale, a marine biologist from the Smithsonian Environmental Research Center in southern Maryland. In three seasons in the Antarctic, Neale had gathered considerable data on the way UV-B affects Antarctic phytoplankton. Through exhaustive sampling and experimentation on these microscopic algae, Neale and his colleagues have found that phytoplankton photosynthesis declines by as much as 8.5 percent when the UV index is high and water circulation patterns keep the algae close to the surface. Given that phytoplankton form the base of the food chain, such a decline could have a knock-on effect on zooplankton, krill, and everything else in the Antarctic. How much, nobody knows.

"It's hard to say what an 8 percent drop in phytoplankton productivity would translate to in terms of krill populations or fisheries yields or the number of whales," Neale explained in his "bomb-proof" shipboard laboratory. (Everything, from the heavy spectrometer to the smallest test tube, was bolted, lashed, or duct-taped to a stationary object lest it be smashed to bits as the *Gould* tossed on the violent seas.) Neale thinks these krill-eating animals probably wouldn't decline by any more than the phytoplankton. But there could be threshold effects. Some population of penguins or whales or whatever might be just barely hanging on, and this small decline in their food supply might push them over the edge. On the other hand, the declines may not matter at all. No one has the answers yet. Neale's best guess is that there will be declines of higher organisms that are "significant, but not catastrophic." Like so much Antarctic science, determining the net result will require more analysis, further experiments, more hours on deck in the frigid Southern Ocean.

〜

THAT EVENING, THE *Gould* began scraping through floes and brash ice, slowly working her way down the western side of the Antarctic Peninsula towards Palmer. We spent a day taking samples in a

2-mile-wide section of the Gerlache Straits that remained ice-free. It was like being on a Swiss lake in winter. Mountains rose on both sides, and the straits were plugged at both ends by pack ice, leaving a stretch of mirror-still water in between.

We shared this ephemeral pond with a multitude of Gentoo penguins. Gentoos prefer open water for feeding and they could be seen everywhere: congregated on chunks of ice, swimming in great rafts and looking very much like discolored loons until, in a burst of speed, they began hurdling out of the water like diminutive porpoises. Low, foggy clouds hung much of the time in narrow bands above the coast, revealing the ice floes and rocky summits but not the mountains in between.

The following morning, a thick field of frozen floes slowed the *Gould*'s progress toward Palmer. The ship pushed or split a path through the snowy floes, which closed in again not 50 yards astern. Tiny Adélie penguins were scattered across the snowscape and observed the approaching ship with curiosity, fleeing only when the *Gould* was beginning to pass them by, wings flailing uselessly behind them as they waddled across the uneven ice. Through some divine misprogramming, the penguins invariably fled on a parallel course with the ship at approximately the same speed; they would safely escape only when they became exhausted and began to fall behind the enormous orange intruder.

In such company, the *Gould* approached the tiny blue buildings of Palmer Station, huddled on a rocky spit of land protruding from the base of a towering glacier, insignificant specks on the edge of a vast, unowned continent.

As we punched our way toward the dock, the station crew turned out to greet us—their first direct contact with the outside world in six weeks. I was standing on deck with Langdon Quetin, whom I knew had already spent fourteen research seasons at Palmer, a cumulative total of five years of his life.

"Is it really that much?" he said, shaking his head with a smile. "I'd never done the math."

Then the landscape along the glacial cliffs on the other side of the harbor arrested his attention. "I can't believe it," he said to himself.

I thought at first he was looking back on half a decade of work spent at this remote outpost. But he wasn't.

Langdon was amazed at how the landscape had changed since he left Palmer only eight months before. Or, to be accurate, how much more land there was. "It looks like another island is emerging from under the glacier," he said, pointing to a stony beach peeking out from the base of a 10-story-tall cliff of ice. "Every year a little more land is revealed. Makes you wonder if we're docking at Gamage Point or Gamage Island."

Palmer residents may not have to wait much longer for that answer, concealed for eons beneath a mountain of ice.

☜

PALMER STATION IS THE SMALLEST of America's three year-round Antarctic research stations, and the only one supplied exclusively by sea. The *Gould* is generally the only way in or out of Palmer, and it usually arrives from Chile only every four to six weeks in season. During winter it doesn't come at all. The station is located on a large glaciated island just off the peninsula's western coast, a site chosen for its proximity to the penguin and petrel colonies spread over two dozen rocky islands within 2 miles of the station. But these can only be reached when there is open water. When the sea freezes over, Palmer's fleet of Zodiacs—inflatable open boats popular with yachtsmen, sailors, and scientists—is hauled up by the boathouse to protect them from being crushed by blocks of ice piled here and there by wind and tides. Scientists and station personnel are trapped ashore, sometimes for weeks on end.

At these times, outdoor ventures are confined to "the backyard": the couple of acres of ledge and boulder between the base of the glacier and the tip of Gamage Point where the station is situated. It's a striking landscape: dark, lichen-covered rocks meeting the frozen shores of the Southern Ocean with the glacier looming high above. Being from Maine, I found these glacially carved rocks strangely familiar. The sizes, shapes, and geology all looked right, right down to the rusty orange lichen and occasional clumps of hardy grasses clinging to any rocks far enough from shore not to be scraped clean by passing icebergs. The coast of my home state must have looked something like this ten thousand years ago.

Weather permitting, one can reach another spit of land called Bonaparte Point by crossing a frigid river of glacial outflow on a hand-operated cable car. This is a thick metal platform 3 feet long and half as wide, suspended on a pulley from a thick cable. If alone, you must drag yourself across the abyss with one arm, an exhausting activity that left me aching for days afterward. You must bring a hand-held radio and report to Palmer's radio room on each end of the crossing. This ensures that a rescue team will come looking for you if you're gone too long. It also allows off-duty personnel some entertainment, since your excruciatingly slow progress across the chasm can be viewed from the station lounge with a pair of binoculars. My clumsy crossing afforded considerable amusement. Sitting out of breath 30 feet above a bored crabeater seal, I listened to the incredible silence all around. The few noises there had a mythic quality: the exhalations of the slumbering seal, the tinkle of ice rising with the tide against the shore, an occasional cracking sound from the glacier above.

Then the radio crackled to life: "Go, Colin, Go!" the voice said. The lounge party was encouraging me ashore in time for dinner.

With the rest of the landscape covered by glaciers, the only remaining excursion is to climb to their summit on Andy's flag path. On a clear day the twenty-minute walk to the top of the trail affords a vista of the sort usually reserved for Himalayan climbers. Jagged mountain peaks rise out of the blanket of glaciers that have hidden the landscape underneath for tens of thousands of years. Approaching the coast the glaciers suddenly end in stunning cliffs that fall to the frozen sea. The clean, cold wind blows over the scene, carrying snowy mists off the peaks above. It all looks so vast and ancient, it's hard to imagine it ever changing.

But like many glaciers in this part of the Antarctic, the Marr has been retreating at a surprising rate over the past quarter century. Every year Palmer's "backyard" grows a little larger as more rock emerges from beneath the melting walls of ice behind the station.

Mark Melcon, an Antarctic veteran known as "Commander" in these parts, has had photographs taken of the glacier from the same location every year starting in 1992. The glacier is retreating from the land at a rate of 33 feet (10 meters) per year. Hanging on the staircase leading to the station bar are group photos of each season's winter-over crew. The guys in the 1972 photo—there were only guys then—are dressed like today's teenagers. The glacier behind them appears to extend all the way to the T-5 communications shed; by the time of my 1998 visit, the shed stood 400 yards forward of the glacial tongue.

Looking at such photographs produces a strange sensation, as if a mountain suddenly disintegrated before your eyes.

∞

ON MY ARRIVAL AT PALMER I was assigned to bunk with Peter Duley, one of Fraser's field scientists. I tromped up the prefab metal stairs to the sleeping quarters of the main building, a modest blue prefab structure with the size and character of a small ski lodge. Half the station personnel were housed in the dozen third-story dorm rooms, which sit atop the administrative and communications offices, galley, and pantry and, on the ground floor, the science labs around which Palmer's world turns. The dorm room doors had numbers on them, but over the years these had been replaced by distinctive stickers—an emperor penguin here, a glued-on plastic iguana there, the Grateful Dead bears just next door. Peter, mustached and often smiling, was engaged in fixing his laptop. He bade me welcome, cleared a spot for my luggage and, later, surrendered precious space in a string bag hung out the window in which I refrigerated a couple of newly acquired beers. The *Gould* is a dry ship.

Peter was a fellow Mainer from the wooden boat–building hamlet of Brooklin. He'd studied marine biology at the College of the Atlantic in Bar Harbor and had worked with whales, puffins, terns, and peregrine falcons along the coast. He'd spent time on the obscure islands, bays, and wilderness harbors of my youth, introducing terns to an outer island of Penobscot Bay, working with the puffin colony on Eastern Egg Rock, winter caretaking a Roque Island farmstead way Downeast. He too had seen similarities between the offshore islands of Maine and the Palmer Archipelago: their lichen-covered rocks and treeless interiors.

A few days later, the seas remained frozen solid, forcing the Zodiacs ashore, so the scientists kept to the labs, processing samples and catching up on the literature being published in more northerly places. The wind whistled around the corner lab where Peter and I talked about the Adélies, which could be seen across the harbor on Torgesen Island, huddled like teens at a rock concert.

"Their problems are primarily related to sea ice conditions," Peter began. "There are those direct effects from loss of winter habitat you talked with Bill about. Then there are indirect effects. Around here there's been less ice, which means you have more evaporation off the surface of the water. That means more snow. Adélies and most other *Pygoscelids* use small stones for their nesting material. But they can't get access to either the nesting sites or the nesting material that they need if there's snow on the ground. The birds can hold on to their eggs [inside their bodies] for a while, but if they are unable to gather enough nesting pebbles than they'll just drop them right in the snow. The snow melts and the egg ends up sitting in a pool of water. So that's a big problem.

"Last year we had snow for so long [in spring] that the birds were drawn to anywhere where elephant seals came ashore to rest. The seals would melt the snow underneath them, exposing pebbles. As soon as they'd leave all the penguins went straight to that spot to collect as many rocks as they could, just anything they could find. It was pretty incredible. They carry them in their beaks one at a time, and they'll travel a hell of a long way to get each one—clear across the other side of the island. We'd see some birds going back and forth all day. Others would hang around the nesting site and take advantage of the situation. When another bird wasn't around they'd come over and take his stones," he laughed. "I guess they all got enough stones in the end, but some birds were doing most of the work. The more experienced ones may realize they can expend less energy by stealing. But we also saw younger birds try to use pieces of ice as nesting material. Of course, it all melts when the bird sits on them, so it's not a very successful strategy."

Snow tends to accumulate on the leeward side of topographical features. When the United States and New Zealand built a joint base upwind of an Adélie rookery in East Antarctica, the rookery became buried in snowdrifts and was abandoned until the base was disassembled many years later. At Palmer, the prevailing wind direction during storms is from the northeast, so when there is snow it drifts on the southwest sides of the rocky prominences of area islands. Fraser's team found that in recent years bigger, longer-lasting snow drifts have rendered eighteen colonies extinct. Of the sixty-six active colonies remaining, only eleven stand southwest of a topographical feature. The implication is, of course, that increased snow is reducing the available nesting sites.

Then there's the matter of ticks. There have probably always been a few around—hitchhikers on terns and other migratory seabirds. In recent years, however, warmer winter temperatures are allowing these parasites to survive and breed in the region. Adélies are particularly vulnerable as they do not preen one another. "Affected birds are thick around the neck with ticks," Pete says. "We've captured penguins where their feet are all white and they're very weak. And you push back their feathers and see hundreds and hundreds of ticks. It's especially hard on fasting males during the first nesting period. They've not eaten in a month and all those ticks begin to take a toll." The ticks suck the Adélies dry.

༄

NOBODY KNOWS FOR CERTAIN IF OUR SPECIES is responsible for the warming of the Antarctic Peninsula. It could be the result of a natural process, the latest act in the planet's ten-thousand-year recovery from the last ice age. Many glaciologists support this view.

On the other hand, global warming models from the early 1970s predicted that climatic effects of human greenhouse gas emissions would be felt first and most strongly at the poles. More than two decades ago scientists prophesied that one of the first signs of human-caused climate change would be the collapse of the Antarctic Peninsula's ice sheets. This is exactly what has taken place.

Regardless of the cause, the oceans—and humanity—will feel the effects of a change in polar climates. Populations of fish, penguins, and whales—creatures we value for commercial, aesthetic, or spiritual reasons—may plummet. Deteriorating glaciers and ice shelves will contribute to the destruction of Cocodrie, Louisiana, Caye Caulker, Belize, and the atolls of the Marshall Islands.

And all the while, the West Antarctic Ice Sheet will loom in the background, with the potential—however improbable—to destroy our world civilization in a Genesis-like deluge.

eight
Sea Change

IT WAS AFTER NIGHTFALL WHEN THE *Venizelos* finally left the Bulgarian port of Varna and set course for the Bosporus. In a few hours the 38,000-ton passenger ship would reach the famous straits, completing her weeklong circumnavigation of the Black Sea.

Out onto the ship's deck came a procession of bearded men in dark robes and tall smokestack-shaped hats, followed by a crowd of passengers. Solemnly they gathered at the rail overlooking the dark, brooding surface of the dying sea. These were the leaders of the Orthodox Christian world, the religious successors of the Byzantine Empire that once encompassed all the shores of the Black Sea. Standing side by side were the Patriarchs of Romania, Bulgaria, and Georgia, surrounded by their retainers. In their midst stood the Ecumenical Patriarch Bartholomew, head of the "mother church" in Constantinople, the former Byzantine capital, known to the rest of us as Istanbul.

With great ceremony, Bartholomew stood at the rail and blessed the dying sea below. The Black Sea had gained an unexpected ally.

This was no ordinary voyage. Gathered aboard the *Venizelos* were four hundred scientists, environmentalists, and religious leaders from around the world. They were here on the unusual joint invitation of Patriarch Bartholomew (the titular spiritual leader of the East) and the Commissioner of the European Union, Jacques Santer (the titular secular leader of the West). The goal of the shipboard symposium was even more unusual. The participants had come not only to see the Black Sea's problems up close but to discuss how religion and science might forge a partnership to save its environment.

This was a tall agenda. Science and religion have been at each other's throats in the West since the Pope arrested Galileo, and the leaders of the world's religions have been less than forceful about confronting our planet's emerging ecological catastrophe.

Until recently, that is. Patriarch Bartholomew has moved his church to center stage in environmental matters. In late 1998 he made church history by proclaiming a new class of sins: those against the environment. "For humans to cause species to become extinct and to destroy the biological diversity of God's creation," Bartholomew said, "to degrade the integrity of Earth by causing changes in its climate, by stripping the Earth of its natural forests or destroying its wetlands, for humans to contaminate the Earth's waters, its land, its air and its life, with poisonous substances—these are sins."

I see these developments through secular eyes: Organized religion's moral track record has been no better than that of governments and other large organizations, all of which can be corrupted by institutional self-interest and personal ambition. Nevertheless, Bartholomew's novel environmental advocacy is enormously important. The leader of one of the oldest and most influential political and cultural institutions in the Black Sea region was taking a stand in support of the living planet in general and the Black Sea in particular. His stance is far more likely to influence public opinion in Eastern Europe than a thousand newspaper articles or a multimillion dollar public education campaign. There are about 150 million Orthodox followers in the Black Sea basin, and the long-suppressed churches of Russia, Ukraine, Romania, and other countries are growing fast. A strong stance by their spiritual leaders could very well build pressure on the region's governments to put greater emphasis on environmental protection.

Furthermore, the Patriarch's chief theologian and heir apparent, Metropolitan John of Pergamon, argued for a new partnership between science and religion. Recognizing the profound, global environmental problems of our age, John argued that scientists and religious leaders need one another. While scientists might identify the causes of a problem—the collapse of the Black Sea, for instance—the solutions often require a comprehensive, large-scale, integrated, long-term strategy involving many nations. Marshalling the necessary support is where spirituality and religion come into play.

"The ultimate problems reside in the human heart and mind," John argued at the symposium's end. "The changes needed will only happen if human attitudes change. And what more powerful force for changing attitudes can there be than spirituality and religion?"

The Black Sea, surrounded on three sides by countries with an Orthodox Christian majority, has gained a most powerful advocate. For the rest of the world ocean, the necessary changes in human attitudes will have to come from other sources.

❧

A CHANGE IN THINKING IS A VITAL PREREQUISITE to addressing the deterioration of the oceans. First, people must understand that the oceans are finite and destructible. Wastes dumped and drained into the ocean do not disappear; they are neither economic nor ecological externalities. Likewise, marine fish and animals are not commodities like iron, wheat, or broilers; they are wildlife. Left to the "logic" of the market, marine organisms become "resources" to be efficiently exploited until so few remain that it is no longer profitable to extract more. What is lost often cannot be regained, denying future generations both the economic and ecological benefits afforded by healthy ecosystems. The oceans—and our grandchildren—deserve better.

Second: Human survival is entirely dependent on the health of the biosphere. Our species evolved to survive within environmental conditions created and maintained by communities of living creatures. Plants maintain a breathable atmosphere, bacteria assimilate wastes and cycle nutrients, and so on. The quality of the whole depends on the interrelationship of the parts. But at the dawn of the third millennium we have reached the point where our impacts on the biosphere are not merely local and temporary but global and, in the scale of human lifetimes, permanent. Our appropriation of more and more of the Earth for our own uses has triggered a mass extinction that rivals the five greatest disasters of the Earth's prehistoric past. This "sixth extinction" is an ongoing event that is already affecting the quality and availability of fresh water, soils, food, and other resources. We have shown that we can make major alterations to the ecosystems of regions as large as the Black Sea and Grand Banks; our earthworks can make shorelines recede by tens of miles along the Louisiana coast. Gasses we've emitted into the atmosphere may well have changed the world's climate, altering landscapes as remote as Antarctica and raising sea levels everywhere. These changes in no way endanger the Earth itself; the planet cares little whether the animal biomass in the Black Sea consists primarily of anchovies or jellyfish. But these changes are a danger to us.

Environmentalists will have to reconsider their central strategic tenet: think globally, act locally. The economic and political forces that propel today's environmental crises act on a global scale and those who care about the environment must do so also. There is no local action the Marshallese can take to address the melting of Antarctic glaciers, just as local activists in Belize—even ecosystem management experts as sophisticated as Janet Gibson—are powerless to protect their reefs from over-heated water and increased ultraviolet radiation. The bureaucratic and economic forces that destroyed the Grand Banks acted on a stage that was largely inaccessible to residents of fishing communities in coastal Newfoundland. Cocodrie and other south Louisiana communities stand to be destroyed by the actions of multinational corporations, federal flood control spending, and the notoriously corrupt political culture in Baton Rouge. One of the greatest environmental disasters of our time may result from events in Antarctica, which has no local population at all.

Protecting the local requires acting on regional, national, and global scales. Governments must take the lead. Granted, they are at best clumsy, at worst corrupt and insidious. But only governments have the reach, the regulatory and enforcement authority, and the international negotiating sovereignty to implement the comprehensive, large-scale solutions required to protect the oceans, atmosphere, and climate. They will only do so if the public demands it.

Then the question becomes, what do we need them to do?

❧

NATIONS HAVE TRADITIONALLY ADDRESSED CRISES facing the oceans in an ad hoc manner, by trying to remedy a single, visible symptom while ignoring the underlying causes. There is an almost pathological failure to see the whole picture. If a particular stock of a certain commercially valuable fish species collapses, fisheries managers might reduce fishing quotas on the threatened fish. The intended recovery might not come, however, since the managers compensate the fishermen by increasing the quota on the very species the threatened fish eats. Or because the crash was actually caused by the destruction of the coastal marshlands in which the fish spawned. Or the distinct "fish stock" the managers think they are managing may be part of a stock in a neighboring country, whose failure to also reduce fishing quotas will doom a recovery effort. Global warming may have increased water temperatures just enough to trigger a mass failure of that fish's eggs, or there may have been a rapid explosion of a population of some tiny planktonic creature that considers those eggs a delicacy.

Or maybe it's a combination of all of these things.

The only way to understand and conserve marine ecosystems is to study, manage, and regulate them *as* ecosystems. The oceans are very large and, not surprisingly, so are their ecological communities. In fact, ecologists can and have divided the oceans into large, functional ecological units based on shared bathymetry, hydrography, and productivity. NOAA identified forty-nine such ecological units, based largely on fisheries data. This list has been further refined by Conservation International, a Washington-based environmental group, which added information on water circulation, bottom topography, salinity, and temperature. The resulting list of seventy-four large marine ecosystems (see the Appendix) includes sheltered seas like the Black, Mediterranean, and Yellow; specific open-ocean currents like the Humboldt off South America; and distinct areas of the continental shelf like the Newfoundland banks, or Antarctica as a whole. These represent an excellent starting point for organizing coherent, ecosystem-scale management units.

The realities of international politics raise two obstacles to this approach. The first is that beyond 200 miles from shore the oceans aren't under anybody's jurisdiction and won't be in the foreseeable future. These vast regions generally consist of the waters over deep abyssal plains, the upper layers traversed by herds of giant tuna, whales, dolphins, and other migratory creatures, the depths largely unexplored and inhabited by strange and unusual creatures. Sadly, protection of the creatures that live or migrate through the high seas will remain under the auspices of international treaties and regimes, which are difficult to negotiate or enforce. We must concentrate our attention on the parts of the ocean that are within national jurisdictions. Not because the high seas are unimportant but because both ocean life and the threats to it are concentrated near land.

Under the United Nations Law of the Sea, all coastal nations of the world have sovereignty over the waters and seabed up to 12 miles offshore and near-total jurisdiction over their Exclusive Economic Zone (or EEZ), which extends a full 200 miles from any inhabitable land. This includes tiny island nations like the Marshalls, which has a land surface area of 70 square miles and an EEZ of 750,000 square miles of the central Pacific. In fact, taken together, the nations of the world have jurisdiction over 36 percent of the ocean surface, including 90 percent of fish stocks and virtually all of the continental shelves. That figure increases to more than 50 percent with the addition of Antarctica, which is shared by twenty-six nations under a stable and ecologically sound treaty system that includes all of the Southern Ocean to 60 degrees south. Legally, governments have the authority to manage and regulate the most critical and productive parts of the ocean.

Of the seventy-four large marine ecosystems, twenty-six lie entirely within the EEZ of a single country. This greatly simplifies management, regulation, and enforcement because the laws, institutions, and priorities of only one nation need to be taken into account. For these nations—which include wealthy industrial democracies like the United States, Canada, Australia, New Zealand, and the United Kingdom—the adoption of one coherent management system for those large ecosystems under their jurisdiction must be made a top priority.

Which raises the second problem. Most ecosystems don't correspond to political boundaries, meaning that meaningful management depends on the cooperative efforts of two or more nations. For some ecosystems, this should be feasible. The United States and Canada share the Gulf of Alaska region of the North Pacific, and the current dispute over commercial salmon fishing there clearly points to the need for joint management. In Europe, the advent of the European Union makes possible the coordinated, central management of several such ecosystems, including the Bay of Biscay and the Iberian coast. It is in the political, ecological, and economic interest of all parties to build mechanisms for coordinating multi-state management of shared marine systems.

In some areas, however, this is at best a long-term goal. In the Mediterranean and Caribbean regions the sheer number of state actors involved makes effective cooperation unlikely. Other ecosystems are shared by nations that are hostile to one another, such as the Yellow Sea, whose polluted waves lap on both North and South Korean shores. The U.N. Environment Program has set up regional bodies to promote cooperation among the nations that share these seas, but they have made little progress. Some ecosystems may continue to be managed in an ineffective, ad hoc manner, but many others can and must be placed under sensible, comprehensive, and coordinated conservation regimes.

Different parts of the ocean have different characteristics; therefore each ecosystem management regime must be custom designed. The first and most important goal is to have all relevant agencies and decision makers operating in tandem rather than at cross-purposes. This might best take the form of a single government or inter-government agency with the legal authority to manage that ecosystem. It would likely include representatives of the ministries or agencies responsible for fisheries, shipping, tourism, land use planning, environmental protection, agriculture, energy, and industry. In most countries it would need to involve relevant state or provincial governments, industrial associations, citizen groups, or legislative committees. Such marine ecosystem management authorities would be responsible for the development and oversight of an overall strategy for the ecosystem. Individual agencies would continue to be responsible for the detailed management of various sectors in accordance with the strategy's aims.

To be of any use, a strategy must not simply be a mechanism for maximizing economic benefits, although improving the quality and quantity of marine resources would most certainly be one of its side effects. Our over-riding purpose in future management of the oceans must be to ensure the conservation and maintenance of healthy, natural ecosystems. Thriving, diverse communities of life not only offer the most benefits to humankind, they are in their own right beautiful, inspiring, and simply "good" in the broadest sense of the word. Fostering such communities on a wide scale is not just the best thing to do, it is the right thing to do.

∞

IT'S A DAUNTING TASK TO MANAGE AND COORDINATE human activities over enormous swathes of coast and ocean. Even with all its wealth and advantages, the United States can't even manage its own education or health care systems. It's fair to ask if any country can hope to manage something as large, complex, and poorly understood as a marine ecosystem. But when they sit down and look at what's at stake, most coastal nations will probably realize that they have to try; the economic and social consequences of letting the oceans deteriorate are simply too great for such nations to ignore.

Fortunately we aren't starting from scratch. Important tools have been developed in recent decades that can be employed to conserve and manage large marine ecosystems. It's worth saying a few words about three of the most promising.

Given that many critical threats to marine life come from the land, marine ecosystem management will necessarily include coastal zones. Governments are finding that an integrated coastal management plan is an essential part of effective marine conservation. The idea is to coordinate policies that affect coastal regions, from fishing and tourism to agriculture, industry, and urban planning. But the specifics will look different from place to place. Belize's effort to set up a single coastal management authority for the entire country is a logical plan for a small nation whose most valuable marine resources are most threatened by land-based pollution and development projects. In another environment, the coastal zone approach might be less important. The Newfoundland Banks might be managed successfully without an integrated land use strategy, as long as new developments take marine impacts into account. Other regions might benefit from strong, proactive efforts to protect wetlands or mangrove forests or to restore those already lost. The specific goals depend on local conditions.

In some regions, ocean life is damaged by pollution originating hundreds or thousands of miles from shore and delivered to the sea by rivers and streams. For these places, an integrated watershed management regime will probably be the only effective strategy. From a marine perspective, a watershed strategy should take steps to limit river-borne pollution and contaminants to levels acceptable to marine life. This might involve the construction of artificial wetlands and wastewater treatment plants, the planting of wooded strips between riverbanks and crop fields, or the introduction of more efficient practices or technologies for industry and agriculture. As with coastal zone management, watershed management might take place within the marine authority (where the actors are largely the same) or under a parallel structure (where they are not). In either case, to succeed a watershed strategy must be comprehensive and coordinated. An ad hoc approach like that currently used in the Mississippi watershed is doomed to be ineffective. Many watersheds are international and will have to make do with regional management regimes like that set up for the Danube.

In many parts of the ocean, vital ecological activities concentrate in specific locations. Fish and other marine life may gather on a specific part of the seafloor at certain times of the year or of their life cycles to spawn, mature, or feed. These unique areas are the biological factories for the surrounding ocean. Bottlenose whales and a host of other marine species congregate to breed in a deep underwater canyon called The Gully at the edge of the continental shelf off Nova Scotia. The shallow sea-grass meadows around Swallow Caye, Belize, are among the few remaining habitats of the manatee and are vital for the health of surrounding reefs.

Just as on land, there are some undersea places that are unique world treasures that ought to be protected for future generations to explore and enjoy. Australia's Great Barrier Reef, the giant kelp forests off Monterey, California, and the deep volcanic vents of the Pacific are the marine equivalent of Yellowstone, Yosemite, or the Grand Canyon.

For these reasons many countries have been experimenting with Marine Protected Areas. These take many forms. Some are the undersea equivalent of wildlife sanctuaries, with all types of fishing and industrial activities prohibited to prevent disruption to a sensitive or highly productive area. Others are more like national parks and monuments, open to regulated use by snorkelers, divers, and even sport fishermen or (where appropriate) commercial activity. Most are zoned for multiple uses, with different activities or types of fishing allowed in each area, but usually under an umbrella of strong legal protections against industrial activity such as pollution, seafloor mining, or offshore energy projects. Many extend onto the land to include mangroves, wetlands, or seabird nesting sites. Whatever the form, they provide vital ecological anchors for the surrounding ocean.

Countries that have established Marine Protected Areas have gained extensive benefits. From New Zealand to Kenya and from South Africa to the Philippines, protected areas have been found to have more and larger fish than unprotected ones. At a minimum, this keeps stressed species from collapsing, particularly species that tend to stay in one area. More often, commercial fishing improves in areas surrounding the reserve, improving the health of industry and ecology alike. Protected areas can also provide scientists with critical baseline ecological information for use in the management or restoration of surrounding areas.

But unless they are embedded into a wider ecosystem management scheme, Marine Protected Areas offer little hope of reversing the deterioration of the oceans. Stresses from surrounding areas—a change in species structure, increased pollution or sedimentation, or the spread of algae blooms and diseases—can doom the protected areas as well. At Hol Chan in Belize, managers can't protect the reserve's reefs from being smothered by sediments and algae blooms triggered by real estate development on surrounding islands. An ecosystem-wide management plan might have prevented the damage by coordinating tourism and conservation objectives over the entire area. In the long term this benefits all parties. A resort on Caye Chapel or Ambergris Caye is far more valuable if Hol Chan's reefs, the area's most popular tourist draw, remain healthy.

A final point about managing marine ecosystems: The central and overriding priority is to keep the ecosystem healthy. In the end the economy is a fully owned subsidiary of the environment, and this is particularly true of ocean-related industries. It is not a question of choosing between the health of the fishing industry and that of the marine ecosystem. In the long term, a healthy fishing industry can only exist if the ecosystem that supports it is also healthy. Conservation may not serve the short-term interests of a specific company or individual, but it promotes the long-term health of entire industries and communities. Focusing on exploitation runs the risk of permanently destroying what should be an everlasting resource.

❧

NOWHERE IS THE TENSION BETWEEN conservation and exploitation as pronounced as in fisheries management. As the only major industry that directly relies on the extraction of marine creatures, fishing has the largest stake in the health of the oceans. That is why it is so vital that fishermen play a central role in large-scale ecosystem management. They are not only the most interested parties but are in many ways the most knowledgeable. It is my hope that the wisdom and support of individual fishermen will be harnessed in the service of a better stewardship of the undersea realm.

In the developed world, individual fishermen are becoming increasingly rare. This is a great tragedy for life on both sides of the surf. Like the farmers a generation ago, the self-employed fisherman of popular imagination is being replaced by industrial-scale operators. The near-shore fisherman and the factory-freezer trawler have no more in common than a family farm has with a modern industrial feed lot. As individual fishermen become rarer, coastal communities continue to lose their ties to the sea. The fishing lobby increasingly represents not men and women of the sea but anonymous, land-bound shareholders, people who have no knowledge of or stake in the long-term health of any particular bay or fishing bank. Large corporations usually make poor stewards. Their executives are rewarded based on their success in maximizing efficiency and short-term profits; this is great if you're manufacturing widgets, but a recipe for disaster when you're dealing with the hunting and gathering of wildlife. Our fisheries are already too efficient for their own good.

Throughout much of the developing world, coastal communities continue to engage in artisanal fishing. Using hand-held spears and throw nets, small, motor-less boats, baited hooks, hand-set nets, or simple traps, several million artisanal fishers catch over 8 million tons of fish a year, 10 percent of the world total. This catch is usually consumed close to the port of landing, often almost entirely by the fisherman's extended family. In tropical developing countries, artisanal fishermen account for between 40 and 100 percent of animal protein consumed by the population. These fisheries rely heavily on the skill and knowledge passed down from generation to generation, a treasure-trove of ecological information that often exceeds that collected by scientists.

All over the developing world, artisanal fishers are increasingly finding themselves competing against large, industrial fishing fleets. In 1997, the West African nation of Senegal signed a four-year agreement with the European Union that allowed large European fishing vessels to operate in near-shore waters used by traditional fishermen. The Senegalese government favors large foreign trawlers because of the access fees they pay, which reportedly account for 70 percent of the country's foreign exchange earnings. Senegal's fifty thousand small-scale fishermen are not so pleased. They blame the large, wasteful European trawlers for the rapid decline of the fish their communities have always relied on for basic nutrition. In Bahrain, foreign shrimp trawlers are accused of damaging coral reefs, damaging fish populations, and uprooting traditional fishing traps. In India, home to 7 million small-scale fishermen, opposition to foreign shrimp trawlers was so great that the government revoked the foreign vessels' licenses. Where large-scale vessels are allowed to fish waters relied on by artisanal fishing communities, the communities—and marine life—wind up much the worse.

The difference is not just technological: It is not just that both groups are equally rapacious and the industrial trawlers are more efficient at it. Their incentives are different as well. In the developed world, individual fishermen and traditional fishing communities are often pushed out of the fishery by large, corporate-owned vessels after local fish resources are decimated by the combined fishing pressures. When stocks collapse, the large mobile fleets relocate to other coasts or are sold to developing countries by their owners. Local fishermen are left with a damaged sea and bleak futures.

In Maine and Atlantic Canada some two thousand coastal villages and island communities have depended on fishing since the day they were founded. Fishing is more than an occupation here: It is the heart and soul of these communities, the basis of local culture and social relations, and the economic lifeblood of virtually everyone living there. Fishing villages have survived in considerable numbers in Ireland, Scandinavia, Iceland, Japan, and other developed nations. Most are still dominated by small, individually owned inshore vessels of less than 45 feet in length, fishing on traditional grounds within 12 miles of shore. Typically they are profitable but labor intensive, thus undesirable in the eyes of more short-sighted economists.

Now that many stocks have collapsed, economists and politicians argue that traditional fishing communities are unsustainable, uncompetitive, and anachronistic. From Newfoundland to New Zealand, we are told there are too many fishermen chasing too few fish, so the small, "inefficient" fishermen must make way for an elite cadre of larger, capital-intensive, highly profitable fishing enterprises. The right to fish the sea should be privatized, and those rights should be bought and sold on the free market, thus ensuring that they become concentrated in the most capable hands. This pleases economists and corporate strategists since it will reduce the number of fishermen while increasing the industry's profitability. With all those small-scale fishermen out of the way, the stocks will be better able to support industrial-scale fishing.

This analysis is both wrong-headed and morally bankrupt. The central problem with the fishing industry is not the number of fishermen but the overwhelming over-capacity of the world's fishing fleet. Using data from Lloyd's Maritime Services, two former senior managers at the United Nations Food and Agriculture Organization (FAO) calculated in 1998 that the fishing industry had twice the capacity needed to catch all available fish. This means that if fishing capacity were cut by 50 percent, the industry would still land just as many fish. So where to make the cuts?

Industrial-scale fishing operations have played a central role in the destruction of fish populations since they are too large, effective, and wasteful for nature to support. UN-FAO statistics show that large-scale vessels (those at least 79 feet long and 100 gross register tons) make up 1 percent of the world's 3.5 million fishing vessels but account for about half of the worldwide catch.

Phasing out the largest vessels would reduce capacity and ecological damage with a minimum of social and economic disruption. It would also be a triumph for free-market competition. Large vessels receive the bulk of the estimated $20–50 billion in government subsidies given to the fishing industry each year, and most could not turn a profit without them. Removing these market distortions would do more good, and far less harm, than throwing all the world's small-scale fishermen out of the profession.

Second, those fishing villages that still exist have more right than anyone to fish their traditional waters. Such fishermen have an intimate knowledge of their fishing grounds, and most have a deep respect for the sea and the creatures that live within it. Their communities grew in symbiosis with the surrounding ocean, towns and villages whose fate is tied to the health of the marine life near its shores. While far from perfect, their conservation and management record is far better than that of the industrial fisheries and the bureaucrats who regulate them. Rather than disenfranchising traditional fishing communities, governments should empower them.

This is not to say that artisanal and small-scale fishing can't cause harm to the marine environment—it can and does, and careful management of these fisheries is essential. But wherever appropriate, coastal fisheries should be placed under a community management framework. In this system, traditional fishing communities are granted proprietary rights to harvest marine life in local waters. Ownership of these rights is vested not with individuals but in the local community in the form of a fisheries trust, so they remain a public resource. Responsibility for the resource would be in the hands of those directly dependent on its long-term health. The community would have collective responsibility for the allocation of fishing rights, ensuring sustainable harvest, and deciding how potential new entrants can join the fishery. Fishing licenses might be granted on a life-long basis, reverting to the community when the fisherman retires or perhaps transferred to his or her children if they wish to join the fishery. The community would set eligibility requirements, compile catch records, and play a central role in the management of the local fishery.

Community fishing trusts could act as the local agents to implement ecosystem-wide conservation or management targets, set local zoning and land use permits, and exchange information with scientists on the ecosystem's health. Assuming they comply with overall ecosystem needs, communities could set their own gear, quota, or season restrictions to ensure that conservation measures take local factors into account. It is a system that would have legitimacy in the eyes of the harvesters since they themselves would play a central role within it.

Placed in a cooperative framework, scientists and fishermen usually find they have a great deal to learn from one another. Scientists often have the most to gain. As Nancy Rabalais of the Louisiana Universities Marine Consortium puts it: "You spend ten years of your life researching some aspect of marine ecology and mention your findings to a fishermen, and half the time they say, 'yeah, I knew that.'" Fishermen have accumulated generations of complex information about the behavior and interactions of the creatures they fish. This knowledge should be harnessed in the service of ecologically sound management, not extinguished in homage to false notions of economic efficiency.

∞

COMMUNITY-BASED MANAGEMENT WON'T WORK in fisheries that are either far from land or controlled by communities where small-scale fishermen are already extinct. In these waters, ecosystem managers will probably have to take a more traditional, top-down approach to fisheries regulation and management. To be effective, these must combine effective monitoring and enforcement with strict limits on total catch, bycatch, gear types, and seasons of operation to ensure that harvests are sustainable. Vessels that cannot fish in a way that is both profitable and ecologically sound should be allowed to fail. If fishing rights are to be privatized as part of a transferable quota scheme, there should be strict limits on the percentage of the total catch any one fishing firm can acquire.

Some fishing practices are so harmful they should be eliminated altogether. Large-scale drift nets were banned globally in 1989 following worldwide outrage over their wanton destruction of marine life. Factory-freezer trawlers should share the same fate; they have caused far too much damage too quickly to too many ecosystems. Smaller drift-nets and bottom trawlers can cause widespread damage and must be carefully regulated where they are used.

Governments must also end the counterproductive practice of subsidizing the construction and operation of fishing vessels. The world has twice as much fishing capacity as it can use, yet governments subsidize their fleets to the tune of $20–$50 billion annually. These resources would be better spent funding marine research and the creation of large ecosystem management mechanisms; even burning the money would make more sense than using it to hasten the destruction of the fish populations the industry relies on.

❧

WHEN I VISITED ST. ANDREWS, NEW BRUNSWICK, in the summer of 1997, the region appeared to be in the midst of a seafood renaissance. As elsewhere in Atlantic Canada, southwestern New Brunswick had been reeling from the collapse of cod, halibut, and other groundfish. But Charlotte County had something Newfoundland and southern Nova Scotia did not: deep, well-protected coves and bays flushed and rejuvenated by the massive, powerful tides flowing in and out of the Bay of Fundy. It appeared to be the perfect place for a new industry: the farming of salmon. The rocky coves would shelter the floating salmon cages from all but the most powerful storms. Fundy tides—the world's largest at 50 feet or more—would carry fresh, nutrient-rich waters into these coves twice daily, flushing all wastes offshore on their retreat. Rather than hunting fish in the sea, New Brunswickers could grow them. Since the first experimental cages were deployed in 1978, salmon aquaculture in Charlotte County had grown to become a $100 million a year industry. Seventy-four salmon farms, some housing as many as 250,000 fish, accounted for more than a third of the county's employment. Salmon mansions sprang up on the shoreline and new businesses catered to the needs of fish farmers. For a region reeling from the collapse of the groundfisheries it seemed an economic miracle.

When I returned a year later the bubble had burst. An outbreak of infectious salmon anemia, a virulent, untreatable fish disease that dissolves the kidneys of salmon and herring, had forced the slaughter of more than 1.2 million farmed fish at dozens of sites. The government announced a $10 million bail-out package and a mandatory fallowing of a quarter of the industry's capacity. Total costs to fish farmers were estimated at at least $40 million. The costs to wild salmon and the wider environment have not been calculated. "There are obviously major problems with salmon aquaculture," I was told by Frederick Whoriskey of the Atlantic Salmon Federation, a St. Andrews–based conservation group that had once championed salmon farming. Salmon farms, he said, needed to "clean up their act and function sustainably."

New Brunswick was discovering that aquaculture isn't always the conservation solution it's cracked up to be.

Some Charlotte County residents had been concerned about the environmental effects of the salmon cages for many years. As the industry expanded, people noticed greasy coatings on the area's rocky shores, larger and more frequent algae blooms, and smelly residues on clam flats. In 1990 a New Brunswick conservation group released a report warning that uneaten salmon feed and fish feces were triggering dangerous algae blooms, destroying bottom life near the cages, and damaging wild salmon stocks through disease and escaped farmed salmon (which carry genetic traits that are unhelpful to wild fish).

Scientific studies released over the next few years suggested serious environmental problems. Bottom life was severely disrupted in the vicinity of the salmon cages, and nutrient pollution was causing algae blooms and hypoxic zones. There were frequent outbreaks of diseases and parasites in the crowded salmon pens, which salmon farmers treated by feeding or bathing the fish in antibiotics and pesticides, leading to widespread releases of potentially toxic substances into coastal waters. The complex tidal currents in the area often carried the farm's pollution not out to sea but onto shore.

A study by the Atlantic Salmon Federation showed that the proportion of farmed salmon in local wild salmon runs increased from

5.5 percent in 1983 to 90 percent in 1994. Wild salmon migrate back to the streams where they hatched, and each population is genetically adapted to a particular stream; farmed salmon do not make these spawning migrations, and the ability of escaped farm fish to displace wild stocks was alarming to ASF. Local fishermen didn't like the farms either, claiming the salmon cages were blocking inshore herring runs, wiping out lobster grounds, and tainting area shellfish beds.

Aquaculture was causing more problems for southwestern New Brunswick than it had solved.

As commercial fish stocks have collapsed, many governments have looked to aquaculture to make up shortfalls of both employment and fish. Global aquaculture production doubled between 1986 and 1996 and at this writing the FAO estimates that it accounts for one-quarter of the world's food fish supply. The industry produces an estimated $43 billion worth of cultured fish and aquatic plants worldwide, and that figure is projected to grow. Farming accounts for about 20 percent of fish production, including 40 percent of mollusks and salmon, and half of all shrimp. Small-scale operations are experimenting with the farming of cod, flounders, and halibut.

At the end of the 1990s the vast majority of the world aquaculture industry farms with little impact on the environment. Freshwater carp, catfish, and tilapia are plant-eaters and are usually raised in special ponds, where they help convert potentially harmful organic wastes into edible fish meat. In China, sustainable aquaculture has been practiced for thousands of years, usually by combining fish ponds with crop and livestock farming in a sort of ecological loop; carp consumed organic farm wastes and were eaten by the farmers, who regularly scraped aquatic plants from the pond's bottoms to use as fertilizer. Shellfish such as mussels, scallops, and oysters filter algae and plankton from sea water, so farming them can be beneficial to the surrounding marine environment.

Unfortunately, much of the industry's recent growth comes from operations with extremely destructive side effects. Shrimp, for instance, grow best in coastal estuaries with a mixed flow of salt and fresh water. As a result, fish farmers often carve shrimp ponds out of mangrove forests, the nursery on which most surrounding marine life depends. Half of Thailand's mangrove forests were cut down during the 1980s, most to make way for shrimp ponds. According to the Worldwatch Institute the shrimp farms raised 120,000 tons of shrimp a year but caused an annual loss of 800,000 tons of harvestable wild fish. Most ponds are abandoned within a decade because of disease outbreaks or excessive contamination and clogging from shrimp wastes.

Most high-value aquaculture operations farm carnivorous species such as salmon, trout, eels, and shrimp. These animals are fed pellets and oils made from small, edible schooling fish like mackerel, capelin, sardines, and anchovies. This has two implications. First, such farming reduces human food supplies since the farms are net consumers rather than producers of fish protein. It requires three to five pounds of wild fish to raise one pound of farmed salmon. Second, these practices undermine the recovery of wild fish populations by creating a ready market not only for small schooling fish but just about any fish, large or small, that can be ground into meal. Currently 6 million tons of wild ocean fish are fed to farmed fish every year, reducing food supplies available to poor people in developing countries. "On land we grow herbivores like chicken and cattle since it's an efficient way to make protein," says Rebecca Goldburg, a staff scientist at Environmental Defense Fund and author of a study critical of such practices. "It makes no more sense to grow carnivores in fish farms than it does to grow tigers on land."

If aquaculture is to contribute to either food security or the recovery of the world's oceans, the industry must move away from such wasteful practices. Fish farmers should be encouraged to raise herbivorous species like carp, tilapia, and shellfish. They should raise them in a manner that does not lead to the pollution of the surrounding environment or the destruction of coastal habitat. If that requires that a species be farmed in fully enclosed inland farms with full wastewater treatment facilities, then that is the way it must occur. If such practices prove unprofitable, then the raising of that species will have to be abandoned. Externalizing production and pollution costs is a most destructive form of market distortion.

✍

THE PROBLEMS ASSOCIATED with global atmospheric change cannot be solved through the large ecosystem management model outlined above, although their effects can be mitigated by improving the health and resilience of marine ecosystems. Reducing the concentrations of greenhouse gases and CFCs in the atmosphere requires a separate set of actions conducted not at the scale of large ecosystems, but at the global level.

The causes of ozone depletion have been largely addressed. Satellite images of the expanding ozone hole over Antarctica prompted over 150 of the world's governments to negotiate and sign the 1987 Montreal Protocol, a legally binding convention that imposes a phased ban on the production of CFCs and other ozone-depleting substances. Later amendments imposed a total ban on CFC production beginning in 1996.

Even with the ban in effect, ozone depletion will get worse before it gets better. CFCs continue to be used in older air conditioners and refrigerators; once released, they reside in the atmosphere for long periods. Chlorine from CFCs is expected to continue accumulating in the upper atmosphere for another decade, and ozone levels over the Antarctic may not fully recover until the second half of the twenty-first century.

For now we must simply wait and stay the course. Individuals, companies, and national governments must comply with their treaty obligations. Older refrigerators and air conditioners must be disposed of properly so that CFCs inside do not escape into the atmosphere. New high-altitude aircraft and space-bound rockets should be designed to reduce their emissions of nitrogen oxides, which also have an impact on the ozone layer. And one day your children and grandchildren will live in a world with proper radiation shielding, perhaps even sharing it with bountiful krill, penguins, coral reefs, and tropical fish.

✍

THE PROGNOSIS FOR REDUCING greenhouse gas emissions is not nearly so good. The political forces resisting such reductions are far more powerful than those opposing CFCs, and although there is considerable scientific evidence linking these emissions to global warming, there is little chance of acquiring certain proof before it's too late to take meaningful action.

For all we know, it may already be too late. A hundred years of industrial emissions may have already set in motion a chain of events that will reorder the Earth's climate. Or these emissions may have contributed only minimally to a long-term cycle driven mainly by non-human factors. The recent collapse of the Antarctic Peninsula's glaciers and ice shelves may be the slow-motion response to warming that occurred years ago, perhaps (or perhaps not) triggered by gases emitted decades ago by automobiles and factories. If we stopped burning fossil fuels altogether, ice shelves and glaciers would probably continue retreating for many years due to the long delays between changes in atmospheric chemistry and the quality of the climate. The oceans will continue to warm and expand, the sea will continue creeping upwards. If we are truly unfortunate the pattern of ocean circulation will change and paradoxically initiate a new ice age.

Or the world's temperature regulation could be very resilient. If we reduce greenhouse gas emissions now, compensatory mechanisms might slow the warming. During this new century sea-rise and glacial retreat might slow. The Marshallese won't lose their homeland, tens of millions will be spared in the world's low-lying flood plains, and the good times can roll on in New Orleans.

In politics, business, and war, few decisions are made with access to complete and certain information. In the absence of certainty we weigh the expected costs of acting against the probable costs of doing nothing. With global warming the costs of not acting are potentially catastrophic, while the investments required to reduce greenhouse gas emissions will make for a more efficient economy, a safer, healthier society, and a cleaner environment regardless of its influence on climate. We can and should act to reduce greenhouse gas emissions, and given the potential consequences of inaction it is grossly irresponsible not to do so.

Unfortunately, efforts to create a binding treaty to reduce greenhouse gas emissions have not been successful, and the prognosis isn't good. The 1997 Kyoto summit resulted in most industrial nations pledging to make modest, non-binding emissions cuts. Large developing countries refused to take measures that might retard economic growth, and the Marshall Islands and other small island nations were left yelling into the wind. Even if America ratifies the treaty, the agreed reductions by themselves will do little to affect greenhouse gas concentrations in the environment. Some scientists think the world will ultimately have to cut emissions by more than ten times as much. Kyoto's targets represent only a first step, valuable as a precedent and catalyst for further change.

Ultimately, addressing global warming will require a change in the type of energy we use. We must shift from coal and oil to cleaner-burning natural gas while phasing in renewable energy sources that do not produce greenhouse gases. The technologies are available, we just need to find the political will to make the necessary investments.

This energy shift doesn't represent a threat to America's consumptive way of life, wasteful and numbing though it may be. People can drive their cars to work, they'll just be powered with a different fuel that might make city air more pleasant to breathe. Suburban homes can be heated in winter and air conditioned in summer and remain crammed full of electric and electronic tools and appliances, but the heat might come from passive solar panels, the electricity from solar-electric cells. Factories and industry will continue producing, only with lower costs since their waste heat is used to generate electricity.

Wind power is currently one of the world's fastest growing industries, tripling its capacity between 1996 and the end of 1998; it's now a $2 billion industry that generates over 8 percent of Denmark's electricity and 11 percent of the electricity needs of the German state of Schleswig-Holstein. According to the Worldwatch Institute, sales of photovoltaic cells—tiny panels that turn the sun's energy directly into electricity—grew by an average of 16 percent per year during the 1990s. The United States is the world's leading producer of these cells, which are used to generate electricity for individual buildings, remote installations, and small electronic devices. Photovoltaic cells can be directly integrated into roofing shingles and window glass. Others look to hydrogen fuel cells, the energy source developed for the U.S. space industry's remote probes; these cells create heat and electricity with a single byproduct: water.

Use of these technologies has grown rapidly, creating increasingly efficient economies of scale in the industries that produce and use them, since governments have subsidized their introduction. Journalist Ross Gelbspan has pointed out that the American government could foster the growth of alternative energy and reduce greenhouse gas emissions simply by diverting the more than $20 billion in annual tax credits and subsidies it provides to the fossil fuel industry. If we're to subsidize the energy industry we might as well do it in ways that enhance the planet's environment.

∽

WHILE IN BELIZE I HITCHED A RIDE on a dive boat bound for Lighthouse Reef, a remote offshore atoll that is home to the Blue Hole, one of the world's more famous dive sites. The reef was 50 miles offshore, and much of the trip from Caye Caulker was spent out of sight of land. It was a still, clear morning and the cobalt blue surface of the sea stretched in all directions, as smooth and still as a forest lake.

Far from land the surface of the sea came alive. Whole schools of flying fish leapt from the water on both sides of the boat, gracefully skipping across the surface before popping underneath in a tiny splash. Seabirds soared overhead, observing our craft with almost penguin-like curiosity, first with one eye then with the other, before gliding off toward shoals of hidden fish.

Then the dolphins appeared.

They heard us coming and had set an intercept course for the boat. To get our attention, they launched themselves out of the water in groups of a dozen or more, apparently excited by the presence of the boat. The coxswain slowed the engines a bit and the dolphins gathered around the boat, effortlessly keeping pace, porpoising contentedly at 15 miles an hour. One after another they frolicked in the bow wave, their backs so close you could reach over the side and touch them. Some corkscrewed through the water, exposing white bellies with each rotation. Others darted back and forth beneath the boat, leaping out of the water for a glance at the peculiar creatures riding inside. The group would keep it up for ten minutes or more until the coxswain, mindful of his schedule, eased the engines to full throttle. The dolphins regrouped astern and resumed their southward journey.

Twenty minutes later another pod of dolphins sprang out of the water a few hundred yards off the bow, and the scene resumed.

This happened three more times that day. On our return trip, the divers aglow from the profusion of marine life they'd seen on the reef and around Blue Hole, we had to pass them by. The coxswain, behind schedule and smiling broadly, saluted the breaching dolphins and sped on into the setting sun.

It's supposed to be like this, I reflected. An ocean full of life, of creatures that inspire and enrich human existence, not just as resources but as neighbors, companions, even friends. One day perhaps we'll end our self-imposed exile from the natural world and accept that all species share the same fate, riding together on a great blue ball through the inky darkness of the cosmos, the only ocean that may truly be without end.

appendix ⑥
Large Marine Ecosystems

The following list includes the seventy-four Large Marine Ecosystems (LMEs) identified by a panel of leading marine scientists for Conservation International, a Washington, D.C.–based environmental organization. An LME comprises a distinct, functional ecosystem and therefore represents the most effective unit for the conservation and management of the oceans.

Several LMEs fall entirely within the jurisdiction of a single country. Others are controlled by only two countries. In these cases I have included the country's or countries' names in parentheses.

1. East Bering Sea (United States)
2. Gulf of Alaska (United States/Canada)
3. Californian (United States/Mexico)
4. Sea of Cortes (Mexico)
5. Gulf of Mexico (United States/Mexico)
6. Southeastern U.S. (United States)
7. Northeastern U.S. (United States/Canada)
8. Scotian Shelf (Canada)
9. Newfoundland Shelf (Canada)
10. West Greenland Shelf (Denmark)
11. Hawaiian (United States)
12. Central Caribbean
13. Humboldt Current
14. Patagonian (Argentina)

15. Uruguayan (Uruguay)
16. Brazil Current (Brazil)
17. Northeast Brazil
18. East Greenland Shelf (Denmark)
19. Iceland (Iceland)
20. Barents Sea
21. Norwegian Shelf (Norway)
22. North Sea
23. Baltic Sea
24. Celtic/Biscay
25. Iberian Coastal (Spain/Portugal)
26. Mediterranean
27. Black Sea
28. Canary Islands
29. Gulf of Guinea
30. Benguela
31. Agulhas
32. West Indian Gyre
33. Somali Coastal
34. Arabian Sea
35. Red Sea
36. Bay of Bengal
37. South China Sea
38. Sulu/Celebes (Indonesia/Philippines)
39. Indonesian (Indonesia)
40. Northwest Australia (Australia)
41. Gulf of Carpentaria (Australia)
42. Coral Sea (Australia/Papua New Guinea)
43. New Zealand Shelf (New Zealand)
44. East China Sea
45. Yellow Sea
46. Kuroshio Current
47. Sea of Japan
48. Oyashio Current (Japan/Russia)
49. Sea of Okhotsk (Russia)

50. West Bering Sea (Russia)
51. Faroe Plateau (United Kingdom)
52. Antarctic Shelf
53. Polynesia
54. New Caledonia (France/Vanuatu)
55. Melanesia
56. Micronesia
57. East Pacific Central
58. Peruvian East Pacific (Peru)
59. Galapagos (Ecuador)
60. Bahamanian/Antillean
61. East Madagascar (Madagascar)
62. Southwest Australia (Australia)
63. Southern Ocean
64. East Australia Central (Australia)
65. North Central Pacific Pelagic
66. South Pacific Pelagic
67. Chagos/Insular Indian
68. Indian Pelagic
69. South Atlantic Pelagic
70. North Atlantic Pelagic
71. Sargasso Sea
72. Pacific Vent Systems
73. Arctic
74. Hudson Bay (Canada)

Acknowledgments

*T*HIS BOOK COULD NOT HAVE BEEN WRITTEN without the help of a great many people—family and friends, scientists and strangers, editors and agent. It's tempting to thank them all by name, from childhood friends and far-flung relatives to college professors, professional colleagues and a thousand-and-one interviewees scattered around the world, all those who shaped my mind, values, interests, and ideas. There isn't the space—in book or memory—to name them all, but I'm grateful all the same.

Like so many people, I owe my parents the most. My father shared with me his love of sea and sails starting well before I can remember. My mother shared her unusual awareness of nature, people, and art, informing and inspiring much that I've done in life, this book included.

I'll always be grateful to my agent, Jill Grinberg, and to William Frucht, my editor at Basic Books, for believing in a first-time writer with an ambitious proposal. Their enthusiasm and support for this project will never be forgotten. Thank you both.

Several people gave generously of their time and knowledge, helping shape the content and efficacy of my manuscript or travel. Tundi Agardy, now at Conservation International, did both, alerting me to valuable places, people, and written improvements. In Washington I owe a great debt to Sam Loewenberg, Wendy Bellion, George Irvine, and Julie Lehrman, dear friends and incisive readers whose suggestions greatly improved the text. Shep Smith, a trusted friend since childhood with whom I have shared many maritime adventures, taught me much about diving; his presence in Belize made my undersea ventures safe, easy, and productive.

By advice and example, my friend Adam LeBor of Budapest helped me find and negotiate the difficult trail from newsprint to hardcover; since I stopped living in Budapest he's hosted me more times than I can count. *Köszönöm!* In Bucharest I'm forever indebted to Aurel and Sylvie Stoica, whose decade of warm-hearted guidance and friendship helped me better understand Romania and, by extension, the problems facing the Black Sea. During my years in Eastern Europe I benefited from the kind encouragement of several editors: Paul Desruisseaux at *The Chronicle of Higher Education*, Mark Abel at *The San Francisco Chronicle*, Clay Jones of *The Christian Science Monitor*, and Mike Moore of *The Bulletin of the Atomic Scientists*. Each gave me support and opportunities that nurtured my development as both journalist and writer. At the *Monitor*, thanks also to Jim Bencivenga, Greg Lamb, and Leigh Montgomery.

Several people went out of their way to help me in my travel and research of this book. In Newfoundland, special thanks to Patricia Betts and Richard Haedrich of Memorial University, and to Richard Chisholm, Cynthia Warren, and Donald Paul of Burin. Thanks also to Nancy Rabalais, Len Bahr, and Eugene Turner in Louisiana, fellow Mainer Karl Schatz, Larry Plummer in Washington, Paul Ehrlich at Stanford, Eve Brantley in Alabama, Mark Lazar in New York, and Jonathan Kelsey in Belize, who criss-crossed the country to guide me through Turneffe's corals. Amongst Antarcticans I'm indebted to Peter Duley, Langdon Quetin, William Fraser, Charles Bentley, David Vaughan, and Andy Young, as well as to the National Science Foundation's Antarctic Media Visitors Program, which facilitated my travels there.

Finally, thanks to the oceans and their many living creatures, for times of peace and joyous wonder.

Notes

PREFACE

PAGE

xi on Chernobyl and glasnost see Murray Feshbach and Alfred Friendly, Jr., *Ecocide in the USSR* (New York: Basic Books, 1992), 11–23, 229–250.

xii on the dumping of radioactive wastes in the Kara Sea and other parts of the Soviet Far North: Thomas Nilsen and Nils Bøhmer, *Sources to Radioactive Contamination in Murmansk and Arkhangel'sk Counties* (Oslo: Bellona Foundation, 1994).

CHAPTER I

8 Mamaia built en masse: Constantin C. Giurescu, ed., *Chronological History of Romania* (Bucharest: Editura Enciclopedica Românâ, 1972), 367.

10 the Argonauts' journey is the earliest voyage epic in Western literature and is believed to have taken place in the thirteenth century B.C. Some modern scholars contended for years that the voyage could not have succeeded in breaching the Bosphorus, until an eccentric team built a Bronze Age galley and rowed it from Greece to Soviet Georgia. For the expedition leader's account, see Tim Severin, *The Jason Voyage* (New York: Simon & Schuster, 1985).

10 Athens' dependence on Black Sea trade: Michael Grant, *The Classical Greeks* (New York: Charles Scribner's Sons, 1989), 181–185.

10 caviar as food for poor: Neal Ascherson, *Black Sea* (London: Vintage, 1996), 5.

11 *Zostera* meadows: Aleksandrovich Zenkevitch, *Biology of the Seas of the USSR* (New York: Interscience Publishers, 1963), 440–441.

11 shellfish banks: Ibid., 441–445.

12 *Phyllophora* described: Ibid., 429–430, 445–446.

12 oxygen production: Yuvenaly Zaitsev, "Recent Changes in the trophic structure of the Black Sea," *Fisheries Oceanography* 1 (1992), 180–189. Superdomes: Elliott Norse, President, Marine Conservation Biology Institute, interview by author, Washington/Redmond, November 1997.

12 uses of kelp: Kingsley R. Stern, *Introductory Plant Biology* (Dubuque, Iowa: Wm. C. Brown, 1988), 322.

12 *Szigetköz* biodiversity: János Tardy, President, Hungarian National Authority for Nature Conservation, interview by author, Budapest, December 1992.

12 *Szigetköz's* cleansing ability: Peter Literathy, Director, VITUKI Institute for Water Pollution Control, interview by author, Budapest, 1994.

13 construction problems: *The Guardian*, 17 November 1992; Andor Farkas,Reflex Energy Office, interview by author, Györ, Hungary, 29 January 1993.

13 Hydrostav director Ivan Čarnogursky's political and family ties in *The New Republic*, 21 December 1992.

14 Dunajská Streda from interviews with citizens of Dunajská Streda by author, 10 March 1993.

14 Danube nutrient load: Alexandru S. Bologa, "Eutrophication, Radioactivity, and Biological Long-term Changes on the Romanian Black Sea Shelf" (paper presented at the Black Sea Symposium, September 1997).

14 Danube dams: *Christian Science Monitor*, 27 March 1997, 1.

16 Phytoplankton species changes: Ibid., 1, 5.

16 Tulcea brochure quotes: Gavrila Simion, ed., *Tulcea* [tourist guide] (Bucharest: Arta Grafica, 1990), 10.

17 53,000 tons of oil: Global Environment Facility (GEF), *Black Sea Transboundary Diagnostic Analysis* (Istanbul: United Nations Development Program, August 1997), 74.

18 Ceauşescu's Delta colonization: *New Scientist*, 29 March 1997, 32; *Swiss Review of World Affairs*, 3 March 1997; *Christian Science Monitor*, 6 November 1991, 12.

18 Dneister and Dneiper: Murray Feshbach and Alfred Friendly, Jr., *Ecocide in the USSR* (New York: Basic Books, 1992), 124–126.

18 Ukrainian research ships in Ascherson, *Black Sea*, 264.

19 Bay of Odessa bloom: Ibid., 259.

20 destruction of *Phyllophora* and *Zostera*: Radu Mihnea, "Pollution Problems and Sources" (paper presented at the Black Sea Symposium, September 1997), 5.

20 *Ascherson, Black Sea*, 259.

20 Burgas harbor: Ascherson, *Black Sea*, p. 259. Romanian dinoflagelletes and Bay of Odessa: Mihnea, "Pollution Problems and Sources," 5.

20 *Phyllophora* biome's destruction: Zaitsev, "Recent Changes in the Trophic Structure of the Black Sea."

20 Moon jelly population boom: Yuvenaly P. Zaitsev, "The Black Sea: Status and Challenges" (presentation to the Black Sea in Crisis Symposium, September 1997), 2.

21 one billion tons: *New Scientist*, 11 November 1995.

21 total fish landings and commercial extinctions: GEF, *Black Sea Analysis*, ii; *Financial Times*, 21 December 1994.

22 dolphins and porpoises: GEF, *Black Sea Analysis*, 125; monk seals: *Washington Post*, 20 June 1994, A1; Giant sturgeons: GEF, *Black Sea Analysis*, 115; oysters, crabs, mussels: Ibid., 124; *Phyllophora*: Ibid., 123.

22 mussel filtration: Mihnea, "Pollution Problems and Sources," 7.

22 two million fishermen: *New Scientist*, 11 November 1995, 39.

22 disease deaths: AFP (Agence France Press), 12 July 1995.

22 beach closures, *Izvestia*, 19 August 1990, 4; Radio Moscow home service, 0700 GMT, 12 May 1990; bacterial growth: Yuvenaly Zaitsev, quoted in *Pravda Ukrainy*, 24 May 1990, 1; ten to twenty times: see *Izvestia*, 11 September 1998, p. 1.

22 *Izvestia*, 11 September 1998, 1.

22 World Bank estimates in *Financial Times*, 21 December 1994, 9.

22 on BSEP see: Black Sea Environment Program (BSEP), *Strategic Action Plan for the Rehabilitation and Protection of the Black Sea* (Istanbul: UNDP/BSEP, 1996).

22 Danube program: Colin Woodard, "The Danube: Not Yet Blue," *Ecodecision* (Summer 1995): 52–54; Environmental Program for the Danube River Basin, *Strategic Action Plan for the Danube River Basin 1995–2005* (Vienna: Danube River Program Task Force Secretariat, 1994).

25 *Mnemiopsis* into Mediterranean: *The Independent*, 18 June 1995.

26 Alexandre Meinesz, interview by author, Washington/Nice, 20 May 1999.

26 *Caulerpa* and urchins: *New York Times*, 16 August 1997, 1.

26 efforts to remove *Caulerpa*: *New York Times*, 16 August 1997, 1.

27 Lake Victoria: Chris Bight, *Life Out of Bounds: Bioinvasion in a Borderless World* (New York: Norton, 1998), 86–92.

Chapter 2

30 quote from Sylvia Earle, *Sea Change: A Message of the Oceans* (New York: Fawcett, 1995), 24–25.

31 depth, volume of oceans from William J. Broad, *The Universe Below: Discovering the Secrets of the Deep Sea* (New York: Simon & Schuster, 1997), 43–46; Elliott A. Norse, ed., *Global Marine Biological Diversity: A Strategy for Building Conservation into Decision Making* (Washington, D.C.: Island Press, 1993), 42.

31 for an account of the development and early trials of modern diving gear, see Jacques-Yves Cousteau, *Silent World* (New York: Ballantine, 1977).

32 on conditions and atmospheric development early in Earth's history: James Lovelock, *The Ages of Gaia: A Biography of Our Living Earth* (New York: Norton, 1988).

33 population figures from National Geographic Society, "Millennium in Maps: Population," *National Geographic* (October 1998).

33 implications of growth, see Lester Brown, Gary Gardner, and Brian Halweil, *Beyond Malthus* (New York: Norton/Worldwatch Institute, 1999); Paul R. Ehrlich and Anne H. Ehrlich, *The Population Explosion* (New York: Simon & Schuster, 1990).

33 mass extinctions from: Stuart L. Pimm, "The Biodiversity Crisis: A Status Report" (abstract of unpublished paper); Richard Leakey

and Roger Lewin, *The Sixth Extinction* (New York: Doubleday, 1995), Chapter 13.

35 more than one billion rely on fish: Peter Weber, *Abandoned Seas: Reversing the Decline of the Oceans*, Worldwatch Paper 116 (Washington, D.C.: Worldwatch Institute, November 1993), 9.

37 herring stomach contents from Daniel Hawthorne and Francis Minot, *The Inexhaustible Sea* (New York: Collier, 1970), 88.

39 ancient marine life: *Discover* (September 1993); *Sea Frontiers* (January 1994); see also University of Kansas, www.oceansofkansas.com.

40 Coelacanths: *Science World*, 2 November 1998; *Washington Post*, 11 November 1998; *Cincinnati Enquirer*, 14 November 1998.

40 Coelacanth and Japanese: *The Times* (London), 25 March 1999.

41 deep ocean vent life from William Broad, *The Universe Below*, 100–113; Helena Curtis, *Biology* (New York: Worth Publishers, 1983), 982.

42 vents and life's origins: Broad, *The Universe Below*, 100–113.

42 fisheries statistics, six small fish species from UN-FAO, *State of World Fisheries and Aquaculture 1998* (Rome: FAO, 1998), Part I and Figure 6.

43 whale declines from Weber, *Abandoned Seas*, 31.

43 commercial uses of whales from Danny Elder and John Pernetta, eds., *Atlas of the Oceans* (London: Chancellor Press, 1996), 86.

43 factory trawlers described in Ken Stump and David Batner, *Sinking Fast* (Washington, D.C.: Greenpeace, August 1996).

43 stellar sea lions: Weber, *Abandoned Seas*, 21.

43 twenty-seven million tons of bycatch from UN-FAO, *A Global Assessment of Fisheries By-catch and Discards*, FAO Fisheries Technical Paper No. 339 (Rome: FAO, 1995).

44 "One of three pounds" from *Maclean's*, 5 October 1998, 54.

44 dolphin, porpoise, seabird bycatch from Norse, *Global Marine Biological Diversity*, 93–95.

44 gulf shrimp bycatch in Earle, *Sea Change*, 173.

44 tropical trawls from Weber, *Abandoned Seas*, 20.

44 shark finning: Marine Fish Conservation Network, "Scorecard #7: Sharks" (September 1998); *The Christian Science Monitor*, 8 December 1999, 6.

44 bio-mass fishing: *Maclean's*, 5 October 1998, 53.

44 U.S. wetlands from Vikki Spruill, ed., *Danger at Sea: Our Changing Ocean* (Washington, D.C.: SeaWeb, 1998), 7.

45 mangroves in Philippines in Norse, *Global Marine Biological Diversity*, 108.

45 4.7 million tons fish lost from World Resources Institute, *World Resources 1992–93* (New York: Oxford University Press, 1992).

45 Wilkinson's coral study: *Bioscience* (October 1997); *USA Today Magazine* (May 1993): 62.

45 Tributylin, see Anne Platt McGinn, *Safeguarding the Health of the Oceans* (Washington, D.C.: Worldwatch Institute, March 1999), 26–27.

47 Arctic contaminants: Ibid.; *Maclean's*, 5 October 1998, 54.

47 NRDC on alien species in ballast water: *Chicago Tribune*, 13 October 1996, 1.

48 green crab invasions see Smithsonian Environmental Research Center website at http://www.serc.si.edu/invasions/green.htm.

49 San Francisco Bay: Chris Bight, *Life Out of Bounds: Bioinvasion in a Borderless World* (New York: Norton/Worldwatch, 1998), 151–152.

49 mean temperature increases: IPCC, *Climate Change 1995: The Science of Climate Change – Summary for Policy Makers and Technical Summary of the Working Group I Report* (Cambridge: Cambridge University Press, 1996), 9.

50 fourteen warmest years: Christopher Flaven, "Last Tango in Buenos Aires," *Worldwatch* (November/December 1998): 11.

50 1998 hottest: Lester R. Brown, et al., *Vital Signs 1999: The Environmental Trends That Are Shaping Our Future* (Washington, D.C.: Worldwatch Institute, 1999), 58–59.

50 IPCC quote: IPCC, *Climate Change 1995*, 10.

51 Pacific plankton decline: *Seattle Post-Intelligencer*, 11 July 1998.

51 coral reef mass bleaching: Rafe Pomerance, *Coral Bleaching, Coral Mortality, and Global Climate Change* (report) (Washington, D.C.: Bureau of Oceans and International Environmental and Scientific Affairs, U.S. Department of State, March 5, 1999).

52 ocean circulation and climate: Peter D. Moore, Bill Chaloner, and Phillip Stott, *Global Environmental Change* (Oxford: Blackwell Science, 1996), 92–94.

53 effects of loss of Gulf Stream: *The Atlantic Monthly* (January 1998): 47–64.

53 sea ice and the global conveyor: Keith S. Stowe, *Ocean Science* (New York: John Wiley & Sons, 1979), 299–301; Wallace S. Broecker, "Chaotic Climate," *Scientific American* (November 1995): 62–68.

54 on global warming and ice ages: *The Atlantic Monthly* (January 1998): 47–64.

CHAPTER 3

57 Cabot's real name was Giovanni Caboto. His famous "baskets" account comes to us from a letter to the Duke of Milan sent on 18 December 1497 by the Milanese envoy to London, Raimondo di Soncino, who gathered intelligence on the Venetian sailor's voyage. On Europe's Asia trade in 1500, see R. R. Palmer and Joel Colton, *A History of the Modern World to 1815* (New York: Alfred Knopf, 1983), 105–107, 250, 251.

58 accounts of past fish and shellfish stocks from Farley Mowat, *Sea of Slaughter* (Shelburne, Vt.: Chapters, 1984), 167, 189, 198, 200.

58 herring: Greenpeace, *The Future of Atlantic Herring in New England* (Washington, D.C.: Greenpeace, 1997), 5.

61 highways: Ted Bartlett, *A Century of Service: The Newfoundland Ferry Story* (Moncton, N.B.: Marine Atlantic, 1998), 11.

61 Burgeo and Burin: Patricia Betts, Historian, Memorial University of Newfoundland, interview by author, Washington/Halifax, N.S., 28 April 1999.

62 Schooner Development Corporation, *Zone 16 Overview*, www.ent-net.nf.ca/schooner/zone16ov.htm.

65 on the history of the early cod fishery, see Albert Jensen, *The Cod* (New York: Thomas Crowell, 1972), 67–71, 85–92; Industry Canada, *History of the Northern Cod*, at SCHOOLNET www.stem-net.nf.ca/cod/; and Robert Kunzig, "Twilight of the Cod," *Discover* (April 1995).

66 on the 1634 "Western Charter" and other anti-settlement laws, see St John Chadwick, *Newfoundland: Island into Province* (Cambridge: Cambridge University Press, 1967), 6–7, 8–9.

66 Irish servants: Industry Canada, *History of the Northern Cod* at SCHOOLNET www.stemnet.nf.ca/cod/; *Irish Times* (Dublin) Weekend Supplement, 1 April 1995, 11.

68 resistance to confederation and failures of self-rule: *Maclean's*, 3 April 1989, 16; *Maclean's*, 6 July 1992, 63.

69 Newfoundland's benefits under confederation: Harold Horwood, *Newfoundland* (New York: St. Martin's Press, 1988), 200–201.

69 referendum results: *Maclean's*, 3 April 1989, 16.

69 on Joey Smallwood: *Daily Telegraph* (London), 19 December 1991, 21.

69 only half of households electrified: Melvin Baker, "Rural Electrification in Newfoundland in the 1950s and the origins of the Newfoundland Power Commission," *Newfoundland Studies* 6 (Fall 1990): 190–209.

70 resettlement: Brian C. Bursey, "Resettlement," in *Encyclopedia of Newfoundland and Labrador* (St. John's: Newfoundland Book Publishers, 1981–1994).

71 $2 billion: *Toronto Star*, 21 December 1991, D6.

71 abandoning salt-fish trade, fish sticks: "Fish Plants," in *Encyclopedia of Newfoundland and Labrador*.

71 Canadian fisheries policy and inshore increases: Michael Harris, *Lament for an Ocean* (Toronto: McClelland & Stewart, 1998), 62–69, 70–71.

72 On cod: Kunzig, "Twilight of the Cod."

72 cod stocks and populations from Fisheries Resource Conservation Council, *Building the Bridge*, Report to the Minister of Fisheries and Oceans (Ottawa: DFO, 1996). See also Ransom Myers, Jeffrey Hutchings, and Nicholas J. Barrowman, "Why Do Fish Stocks Collapse? The Example of Cod in Atlantic Canada," *Ecological Applications* 7, no. 1 (1997): 91–106.

73 cod herding and spawning: Jensen, *The Cod*, 22; Kunzig, "Twilight of the Cod."

73 cod eat anything: Jensen, *The Cod*, 27–29; Mark Kurlansky, *Cod: A Biography of the Fish That Vhanged the World* (New York: Penguin, 1997), 33.

73 inshore largely unchanged until WWII: Kurlansky, *Cod*, 127; Harris, *Lament for an Ocean*, 58–59.

74 early post-war fishery: Kurlansky, *Cod*, 127–130; Harris, *Lament for an Ocean*, 52.

74 technological innovations and catch rates 1875–1945: Jeffrey Hutchings and Ransom Myers, "The Biological Collapse of the Atlantic Cod off Newfoundland and Labrador," in Ragnar Arnason and Lawrence Felt, eds., *The North Atlantic Fisheries: Successes, Failures, and Challenges* (Charlottetown, P.E.I.: The Institute of Island Studies, 1995), 39–93.

74 *Fairtry* size and capabilities: William Warner, *Distant Water: The Fate of the North Atlantic Fisherman* (Boston: Little, Brown, 1977), 30–31, 37–44.

75 factory-trawler fleets: Ibid., 50–51; Greenpeace, *Sinking Fast* (Washington, D.C.: Greenpeace, 1986), Chapter 2.

75 Dutch ships converted (the *Willian Barendsz*) Greenpeace, *Sinking Fast*, Chapter 1.

75 size of modern factory-trawlers: Greenpeace USA, "Strip-Mining the Sea" (issue brief) (Washington, D.C.: Greenpeace, 1997); Kunzig, "Twilight of the Cod."

76 size of factory fleet in 1970s: Harris, *Lament for an Ocean*, 56.

76 bycatch rates: UN-FAO, *A Global Assessment of Fisheries By-catch and Discards*, FAO Fisheries Technical Paper No. 339 (Rome: FAO, 1994).

76 pair trawls: Warner, *Distant Water*, 104–105, 120–121.

77 catches peak and fall: Claude A. Bishop and Peter A. Shelton, *A Narrative of NAFO Divs. 2j3KL Cod Assessments from Extension of Jurisdiction to Moratorium*, Canadian Technical Report of Fisheries and Aquatic Sciences 2199 (Ottawa: Minister of Public Works, October 1997), 60.

77 Myers and Hutchings study: Jeffrey Hutchings and Myers, "Biological Collapse of the Atlantic Cod off Newfoundland and Labrador," in Arnason and Felt, eds., *North Atlantic Fisheries*, 44.

78 inshore catch fell: Harris, *Lament for an Ocean*, 65.

79 inshore fishermen figures: Ibid., 69.

79 capelin quota to USSR: Warner, *Distant Water*, 236.

81 Canada's new fleet: *Sierra* (July 1995).

82 By the early 1990s immature fish made up nearly 90 percent of reported northern cod landings, according to Ransom Myers, Jeffrey

Hutchings, and N. J. Barrowman, "Hypotheses for the Decline of Cod in the North Atlantic," *Marine Ecology Progress Series* 138 (1996): 304–305 (especially Figures 2 and 7). Note also that the off-shore fleet's landings, while largely from the so-called Northern Cod stock, also include fish from lesser banks like the St. Pierre Bank near the Burin Peninsula. Cod on these lesser banks also collapsed.

82 DFO overestimates stock: Myers, Hutchings, and Barrowman, "Hypothesis for the Decline of Cod," 304; Harris, *Lament for an Ocean*, 107.

83 DFO's management techniques: Bishop and Shelton, *Cod Assessments*, 3–13.

83 1988 stock assessments and 1989 quota recommendation: Bishop and Shelton, 29–30. The original source document for the recommendation is Canadian Atlantic Fisheries Advisory Committee (CAFSAC), *Groundfish Subcommittee Report 89/1*.

83 1989 quota: Bishop and Shelton, *Cod Assessment*, 60. This was a departure from fisheries science rules (see ibid., 30).

83 1990 quota: Ibid., 60. This quota includes 2,262 tons for France (ibid., 30). The 125,000 ton recommendation was one of two figures forwarded by DFO scientists. A second figure—191,000—was "an interim solution to the problem of going from high" fishing mortality to a sustainable level of 125,000. For more on this, see ibid., 29.

83 1991 catches and quota: Ibid., 60. (See also their footnote 3 for correction to earlier statistics.)

83 January 1992 spawning cod estimated at 130,000 tons: Ibid., 37. Source documents are CAFSAC Advisory Document 92/2 and CAFSAC, *Groundfish Subcommittee Report 92/2* (1992).

83 Ottawa's quota: Bishop and Shelton, p. 37. DFO scientists had recommended a quota of 95,000 tons.

83 collapse of spawning biomass of various zones. Myers, Hutching, and Barrowman, "Why Do Fish Stocks Collapse?," 91–106.

84 Newfoundland population decline: Harris, *Lament for an Ocean*, 246.

86 economic effects of moratorium from Harris: economy shrinks, p. 242–44, 246; public employees fired and MU grants cut, pp. 242–3; Newfoundlanders on the dole, p. 244; population changes, p. 246.

87 lack of stock recovery in various species from Fisheries Resource Conservation Council, *Building the Bridge* (Ottawa: FRCC, October 1996); Department of Fisheries and Oceans, *Overview of the Status of Canadian Managed Groundfish Stocks in the Gulf of St. Lawrence and in the Canadian Atlantic Stock*, Status Report 96/40E (Ottawa: DFO, June 1996).

88 decline of commercial and non-commercial species abundance and size: Richard Haedrich, "Thermodynamics for Marxists: New Ways of Looking at Old Problems" (unpublished paper given at conference, 1998).

88 barndoor skate: Ransom Myers and Jill Casey, "Near Extinction of a Large, Widely Distributed Fish," *Science*, 31 July 1998, 690.

89 Myers quoted in: *Washington Post*, 2 August 1998.

89 trawling effects study: Les Watling and Elliott Norse, "Disturbance of the Seabed by Mobile Fishing Gear: A Comparison to Forest Clearcutting," *Conservation Biology*, 6 (December 1998):1180–1197.

90 catch statistics from Haedrich, "Thermodynamics for Marxists"; DFO Statistics Department tables, posted at www.ncr.dfo/communic/statistics/landings.

91 problems of lobster fishery: Fisheries Resource Conservation Council, *A Conservation Framework for Atlantic Lobster* (Ottawa: Minister of Supply and Services, November 1995).

92 capelin catch statistics: DFO Science, *Newfoundland Region Stock Status Report B–20–02* (1997).

94 Excerpt from "He said, I said," in Jack May, *From Pigs to Politicians* (Twillingate: Landwash Enterprises, 1998), 36.

CHAPTER 4

99 toxic release statistics from Louisiana Department of Environmental Quality, *Louisiana Toxic Releases Inventory Report 1996* (Baton Rouge: DEQ, 1996), Appendix. All statistics are from 1995.

102 jubilees in Mobile Bay: *Science News*, 10 February 1996.

102 Louisiana jubilees in the 1990s: *The Times-Picayune*, 14 August 1996; *The Houston Chronicle*, 31 May 1996.

102 1994 fish kill: *The Times Picayune*, 25 June 1994.

102 fish kills and Hurrican Andrew: *The Houston Chronicle*, 4 September 1992; *Audubon* (September 1995).

107 Descriptions of pre-drainage Iowa are from Calhoun County in 1860, as reported by the *Des Moines Leader*, 1 March 1892.

108 virgin prairie conditions: Douglas R. McManis, *The Initial Utilization of the Illinois Prairies 1815–1840*, Department of Geography Research Paper 94 (Chicago: University of Chicago, 1964), 39, 57; Paul W. Gates, *The Illinois Central Railroad and Its Colonization Work* (New York: Harvard University Press/Johnson Reprint Corp., 1968), 9, 11.

108 tile factories and drainage from Thomas Dahl and Gregory Allord, "Technical Aspects of Wetlands: History of Wetlands in the Conterminous United States," *National Water Summary on Wetland Resources*, US Geological Survey Water Supply Paper 2425 (Denver: USGS, 1998).

108 fifty million acres: Donald A. Goolsby et al., *Flux and Sources of Nutrients in the Mississippi-Atchafalaya River Basin* (report submitted to the White House Office of Science and Technology, Policy Committee on Environment and Natural Resources Hypoxia Work Group) (Denver: US Geological Survey, May 1999), 19.

108 fertilizer use and corn: *Chicago Reader*, 31 July 1998.

108 fertilizer increases: Goolsby et al., *Flux and Sources of Nutrients*, 44 and Figure 5–7.

109 seventy percent of nitrogen from upper Mississippi: Richard B. Alexander, Richard Smith, and Gregory Schwarz, *The Regional Transport of Point and Non-point Source Nitrogen to the Gulf of Mexico* (paper prepared for the US-EPA Gulf of Mexico Hypoxia Management Conference, December 5–6, 1998).

109 nitrogen sources: Donald A. Goolsby et al., *Flux and Sources*, 77 (see p. 108 note above).

109 Iowa's nitrates: *Des Moines Register*, 10 August 1998, 9.

109 ISU study: H. P. Johnson and James L. Baker, *Field-to-Stream Transport of Agricultural Chemicals and Sediment in an Iowa Watershed: Part II. Data Base for Model Testing (1979–1980)*, Report No. EPA-600/S3-84-055 (Athens, Ga.: Environmental Research Laboratory, May 1984); see also William G. Crumpton and James L.

Baker, "Integrating Wetlands into Agricultural Drainage Systems: Predictions of Nitrate Loading and Loss in Wetlands Receiving Agricultural Subsurface Drainage," in J. Kent Mitchell, ed., *Integrated Research Management & Landscape Modification for Environmental Protection* (St. Joseph, Mich.: American Society of Agricultural Engineers, December 1993), 118–126.

109 nitrates in drains: James L. Baker, Iowa State University, interview by author, Washington/Ames, Iowa, 23 March 1999.

110 Raccoon River farms and drinking water: *Des Moines Register*, 23 June 1996.

110 other Iowans drinking unsafe water: *Des Moines Register*, 23 February 1996.

111 history of levees from *The Advocate* (Baton Rouge), 24 April 1994, 6A.

112 development of the Yazoo Delta: John M. Barry, *Rising Tide* (New York: Simon & Schuster, 1997), 96–110.

112 Swamp Act: *The Advocate* (Baton Rouge), 24 April 1994, 6A.

113 The definitive account of the development and result of the Corps's flawed "levees-only" strategy is Barry, *Rising Tide*, especially 39–42, 90–92, 156–159, and 165–166.

113 flood history: Ibid.; *The Advocate* (Baton Rouge), 24 April 1994, 6A.

114 hydrologic history of Delta: Louisiana Coastal Wetlands Conservation and Restoration Task Force and Louisiana Wetlands Conservation and Restoration Authority, *Coast 2050: Towards a Sustainable Coastal Louisiana* (Baton Rouge: Louisiana Department of Natural Resources, 1998), 19–22.

115 hole size and 250,000 tons of rescue boulders: *New York Times*, 15 January 1991.

116 Atchafalaya land gain from *Natural History* (June 1985); and http://www.lacoast.gov/Programs/CWPPRA/Projects/Atchafalaya/index.htm, which includes USGS-NWRC time-lapse images of land loss and gain there.

116 Louisiana land loss: *Coast 2050*, 1, 31.

117 French Quarter submerged: *The Advocate* (Baton Rouge), 2 June 1997.

117 storm surge model: *Coast 2050*, 64.

117 The Superdome was pressed into service when Hurricane Georges struck nearby in 1998. For accounts of this and vertical evacuation

generally, see *The Advocate* (Baton Rouge), 21 November 1998, 1B and 28 September 1998, 1A; *Times-Picayune* (New Orleans), 24 July 1993, A1.

118 Red Cross: Coalition to Restore Coastal Louisiana, *No Time to Lose: Facing the Future of Louisiana and the Crisis of Coastal Wetland Loss* (Baton Rouge: CRCL, 1999), 29–30.

119 gulf fishery value and relationship to estuaries: *Times-Picayune*, 26 March 1996; Coalition to Restore Coastal Louisiana, *No Time to Lose*, 21–25; Gulf Restoration Network, "Facts about the Gulf" (circular) (New Orleans: GRN, 1998).

121 Louisiana governors and environment: Buddy Roemer, *The New Democrat*, September 1991; Edwin Edwards, *The Times-Picayune*, 7 November 1993; Mike Foster, *Dallas Morning News*, 8 November 1998 and *The Progressive*, 1 March 1998.

122 Louisiana's land loss strategy: *Coast 2050.*

123 The bayou country is approximately six thousand years old and encompasses 3 million acres, so was built at an average rate of 500 acres per year. The diversions' promised returns by 2050 are from *Coast 2050*, 122–125.

123 cost/benefit analysis of restoration projects: R. E. Turner and M. E. Boyer, "Mississippi River Diversions, Coastal Wetland Restoration/Creation and an Economy of Scale," *Ecological Engineering* 8 (1997): 117–128.

124 oil and gas canals: *Times-Picayune* (New Orleans), 25 May 1997, p. A1.

124 thirty to forty acres lost: W. H. Connor and J. W. Day, "The Ecology of the Barataria Basin: An Estuarine Profile," *Biological Report* 85 (1987): 7–13.

124 effects of canals: Leanne Lemire, "Backfilling Canals to Restore Coastal Wetlands," in *Restoration and Reclamation Review Vol. 2* (Minneapolis: University of Minnesota, Spring 1998); Department of the Interior, *Impact of Federal Programs on Wetlands, Vol. II* (report to Congress by the Secretary of the Interior) (Washington: DOI, March 1994), Chapter 8.

124 Turner study: R. Eugene Turner, "Wetland Loss in the Northern Gulf of Mexico: Multiple Working Hypotheses," *Estuaries* 20, no. 1 (1997): 1–13; *Times-Picayune* (New Orleans), 25 May 1997, A1.

126 Gulf Program's $12 million: Gulf of Mexico Program, *1997 Share-holder Report* (Stennis, Miss.: Gulf of Mexico Program, December 1997).

127 the watershed strategy: Weeks Bay Watershed Project, *Management Plan for the Weeks Bay Watershed* (Fairhope, Ala.: WBWP, April 1998).

CHAPTER 5

132 Another reason for the collapse of the spiny lobster is the advent of a nearly insatiable foreign market for the hapless crustacean. Starting in 1979, Red Lobster, the American restaurant chain, began buying huge numbers of lobsters from Caye Caulker fishermen. The spiny lobster—unlike the American lobster—can be killed and frozen solid for shipping and storage, a considerable advantage for the chain. See *Americas* (November 1993): 20.

133 Belize's tourists: Belize Tourist Board, *Belize Tourism Statistics 1997* (Belize: BTB, 1998).

133 new Ambergris developments: *San Pedro Sun*, 11 April 1997; *San Pedro Sun*, 9 August 1996.

133 Caye Chapel: *The Courier-Journal* (Louisville, Ky.), 18 August 1993.

134 Belize Audubon Society's role: Osmany Salas, Executive Director, Belize Audubon Society, interview by author, Belize City, 28 September 1998; Valdemar Andrade, Advocacy Coordinator, Belize Audubon Society, interview by author, Washington/Belize City, 12 July 1999.

139 Belize's economic and tourism statistics: World Bank, *Belize Environmental Report*, Report No. 15543-BEL (Washington, D.C.: World Bank, 30 May 1996), 14, Annex VI, 3.

139 Belizean nature protection: Mark Whatmore and Peter Eltringham, *Guatemala and Belize: The Rough Guide* (London: Rough Guides, 1996), 497–500; Lisel Alamilla and Anna D. Hoare, *Coastal Treasures of Belize* (Belize: Angelus Press, 1996); World Bank, *Belize Environmental Report*, Report No. 15543-BEL (Washington, D.C.: World Bank, 30 May 1996), 12, 14, Annex III.

142 nutrient run-off and Hol Chan: World Bank, *Belize Environmental Report*, Annex III, 4; Janet Gibson, United Nations Development Program–Belize, interview by author, Belize City, 29 September 1998.

143 For more information on Belize's Coastal Zone Management, see World Bank, *Belize Environmental Report*, Annex III, 16–24.

149 On the history of animals' conquest of land, see Carl Zimmer, *At the Water's Edge: Macroevolution and the Transformation of Life* (New York: Free Press, 1998).

152 pearlfish: Eugene H. Kaplan, *A Field Guide to Coral Reefs* (Boston: Peterson Field Guides/Houghton-Mifflin, 1982), 249–250.

153 required conditions for coral reefs discussed in detail in Charles R. C. Sheppard, *Natural History of the Coral Reef* (Dorset, UK: Blandford Press, 1983), 5; Sue Wells and Nick Hanna, *The Greenpeace Book of Coral Reefs* (New York: Sterling Publishing, 1992), 20–22.

153 confusion over reef ecology: Herold J. Wiens, *Atoll Environment and Ecology* (New Haven: Yale, 1962), 281–286; Sheppard, *Natural History of the Coral Reef*, 33.

154 coral reef growth: Sheppard, *Natural History of the Coral Reef*, 20–29.

155 Sand 'cements': Ibid., 95–96.

156 clownfish behavior: William Gray, *Coral Reefs and Islands* (Newton, UK: David and Charles, 1993), 75; cleaning stations: Ibid. 72.

156 Caribbean vs. Indo-Pacific: Ibid., 18.

158 Wilkinson coral study: Clive Wilkinson, "Coral reefs of the World Are Facing Widespread Devastation: Can We Prevent This Through Sustainable Management Practices?" in *Proceedings of the 7th International Coral Reef Symposium, Vol. 1* (Guam: University of Guam, 1993), 11–22.

158 Southeast Asia's protein needs according to S. K. T. Yong, "Coastal Resource Management in the ASEAN Region: Problems and Directions," in T. E. Chua and D. Pauly, eds., *Coastal Area Management in Southeast Asia*, ICLARM Contribution No. 543 (Makati City, Phillippines: ICLARM, 1988).

158 Quote and figures on coral cover in Southeast Asia: C.R. Wilkinson et al., "Status of Coral Reefs in Southeast Asia: Threats and

Responses," in *Proceedings of the Colloquium on Global Aspects of Coral Reefs* (Miami: University of Miami, 1993), 311–317.

159 Jamaican declines from Terrence P. Hughes, "Coral Reef Degradation: A Long Term Study of Human and Natural Impacts," in *Proceedings of the Colloquium on Global Aspects of Coral Reefs* (Miami: University of Miami, 1993), 208–213.

160 Jackson's paper: J. B. Jackson, "Reefs Since Columbus," in *Proceedings of the 8th International Coral Reef Symposium* (Panama City: University of Panama/Smithsonian Tropical Reseach Institute, June 1996).

CHAPTER 6

166 demographics: P. Holthus et al., *Vulnerability Assessment of Accelerated Sea Level Rise; Case Study: Majuro Atoll, Marshall Islands* (Apia, Western Samoa: South Pacific Regional Environment Program, December 1992), 44.

167 adult diabetes: *Baltimore Sun*, 26 October 1997, 1A.

167 U.S. compact deal: World Bank, *Pacific Island Economies: Building a Resilient Economic Base for the Twenty-First Century*, Report No. 13830-EAP (Washington, D.C.: World Bank, 8 June 1995), 111. The $1 billion figure is adjusted for inflation.

168 U.S. grants and GDP: United Nations Development Program, "Sustainable Development in the Marshall Islands" (UNDP-Fiji website: www.undp.org.fj).

168 Pell grant funding: See Colin Woodard, "Small Pacific Nations Fear Loss of Pell Grants Could Wipe Out Their Colleges," *The Chronicle of Higher Education*, 24 July 1998, A33.

169 144 homes destroyed: *Baltimore Sun*, 26 October 1997, 15A.

172 inundation of freshwater lens: *Pacific Island Development Report* (PIDP), 19 May 1998; PIDP, 25 May 1998; South Pacific Environment Program, *Background Report: Changing Climate and Sea Levels Affect Pacific Countries* (Apia, Samoa: SPREP, 3 May 1998); Holthus et al., *Vulnerability Assessment*, 26.

172 sea-level rise to date: IPCC, "Changes in Sea Level," in *Climate Change 1995 – The Science of Climate Change: Contribution of Work-*

ing Group I to the Second Assessment Report of the Intergovernmental Panel on Climate Change (Cambridge: Cambridge University Press, 1995), 365–6; *Scientific American Presents* (Fall 1998): 34–35; British Antarctic Survey, "Antarctica: Climate Change and Sea Level" (statement prepared by Ice and Climate Division, October 1998).

173 global warming trends: IPCC, *Climate Change 1995: The Science of Climate Change – Summary for Policy Makers and Technical Summary of the Working Group I Report* (Cambridge: Cambridge University Press, 1996), 31–33; predicted trends: 39.

173 sea-level rise breakdown: IPCC, "Changes in Sea Level," in *Climate Change 1995*, 381–387. Over the past century, the IPCC thermal expansion raised sea levels by 2–7 cm, glaciers by 2–5 cm, Greenland's ice sheet by – 4–4cm, Antarctica – 14–14cm, plus perhaps – 5–7 cm from the net capture or release of surface and ground water (ibid., 380).

173 total sea-rise estimates to 2100: Ibid., 381.

175 IPCC sea-rise study: IPCC, *The Regional Impacts of Climate Change: An Assessment of Vulnerability* (Cambridge: Cambridge University Press, 1998).

175 costs to protect U.S.: Ibid., 304.

176 NOAA-funded study: P. Holthus et al., *Vulnerability Assessment*.

176 A fifty-year storm is not the same as a typhoon. Holthus et al. (16–19) calculated the severity of storm inundation based on data collected by the U.S. Naval Weather Service Command between 1938 and 1970. These data suggest that a fifty-year storm will generate waves with a maximum height of 19 feet. A typhoon event would be more severe, but much less likely: No typhoon has struck either Kwajalein or Majuro in over thirty years and, based on historical data, a direct hit to Majuro would not be statistically expected during a fifty-year period. However, the study notes that global warming could potentially alter storm tracks, creating a "potential typhoon threat at Majuro atoll."

176 author's estimates on costs of protecting Majuro atoll and the Marshalls based on P. Holthus et al., *Vulnerability Assessment*.

176 Quote from Ibid., 69.

177 Marshallese arrival: Marshall Islands Visitors' Authority, *Marshall Islands Visitor's Guide* (Majuro: MIVA, April 1998), 18; Steve Thomas, *The Last Navigator* (Camden, Maine: International Marine, 1997).

178 Bravo test: Jonathan Weisgall, *Operation Crossroads: The Atomic Tests at Bikini Atoll* (Annapolis, Md.: Naval Institute Press, 1994), 302–303.

178 number of U.S. atomic tests: Nuclear Claims Tribunal, *Annual Report to the Nitijela for the Calendar Year 1996*, (Majuro: NCT, 1997), 20–21.

178 AEC and Rongelap contamination: Weisgall, *Operation Crossroads*, 305–306, 313.

179 Rongelapese evacuation and health problems: David Robie, *Eyes of Fire: The Last Voyage of the Greenpeace Warrior* (Philadelphia: New Society Publishers, 1987), 21–29.

179 Rongelapese sent back: Ibid., 29–30.

180 Bikinian exile islands: Jack Niedenthal, *The People of Bikini: From Exodus to Resettlement* (Majuro: Bikini Municipal Government, 1997), 13.

182 Bikini radiological studies: International Atomic Energy Agency, *Radiological Conditions at Bikini Atoll: Prospects for Resettlement*, Radiological Assessment Reports Series (Vienna: IAEA, 1997); Steven L. Simon and James C. Graham, *RMI Radiological Survey of Bikini Atoll*, RMI Nationwide Radiological Survey (Majuro: Government of the Republic of the Marshall Islands, February 1995).

184 results of Kyoto conference: Christopher Flavin, "Global Climate: The Last Tango," *Worldwatch* (November/December 1998): 14.

185 developed, large developing nations at Kyoto: *U.S. News and World Report*, 15 December 1997.

185 Bikenibeu Paeniu, speech at the Fourth Conference of Parties to the Kyoto Climate Change Convention, Buenos Aires, 12 November 1998.

187 examples and maps of FSM sea-rise implications: John Mooteb, *FSM National Government Climate Change Program* (Palikir: FSM Government Palikir, 1997).

CHAPTER 7

194 *Gould* design errors: Antarctic Support Associates, "Summary Report of the Antarctic Research Vessel Committee (ARVOC) Meeting" (Kenner, La., 9–10 October 1997), http://www.asa.org/marine/common/199710mn.htm#1mg status, viewed 13 June 1999.

197 Antarctic Peninsula warming of 5 degrees F (2.5 degrees C): J. C. King, "Recent Climate Variability in the Vicinity of the Antarctic Peninsula," *International Journal of Climatology* 14 (1997): 357–369; Pedro Skvarca, Wolfgang Rack, Helmut Rott, and Teresa Ibarzabal y Donangelo, "Evidence of Recent Climatic Warming on the Eastern Antarctic Peninsula," *Annals of Glaciology* 27 (1998).

197 average temperatures since last ice age: Charles Bentley, geophysicist, University of Wisconsin, communication with author, Washingon/Madison, Wisc., 28 June 1999. At the height of Wisconsin glaciation eighteen thousand years ago, New England was under at least 4,000 feet of ice, covering the White and Green Mountains, while Hudson Bay and the Canadian Shield were more than 2 miles thick, according to Samuel W. Matthews, "Ice on the World," *National Geographic* (January 1987): 91.

197 collapse of ice shelves: David Vaughan and C. Doake, "Recent Atmospheric Warming and Retreat of Ice Shelves on the Antarctic Peninsula," *Nature* 379 (1996): 328–331; Skvarca et al., "Evidence of Climatic Warming"; H. Rott, P. Skvarca, and T. Nagler, "Rapid Collapse of Northern Larsen Ice Shelf, Antarctica," *Science* 271 (1996): 788–792; C. L. Hube, "Recent Changes to Antarctic Peninsula Ice Shelves: What Lessons Have Been Learned?" *Natural Science*, 11 April 1997.

198 temperatures and Larsen-A collapse: Skvarca et al., "Evidence of Climatic Warming"; and Hube, "Recent Changes to Antarctic Ice Shelves."

199 Greenpeace icebreaker: Greenpeace International, "Greenpeace Charts New Antarctic Waters" (Antarctic tour press release, 4 February 1997).

199 dimensions of Larsen-A collapse: Rott et al., "Rapid Collapse of Larsen Ice Shelf"; Skvarca et al., "Evidence of Climatic Warming."

200 1998 iceberg calving: *Washington Post*, 16 October 1998, A3.

200 Filchner station: Alfred-Wegener-Institut fuer Polar und Meereforschung, "Filchner Station," www.awi-bremerhaven.de/Polar/filchner.html, viewed 13 June 1999.

200 Larsen-B cracks: Greenpeace International, "Large Cracks Threaten Collapse of Antarctic Ice Shelf" (Antarctic tour press release, 5 February 1997).

200 Larsen-B and scientists' predictions: National Snow and Ice Data Center, "Satellite Images Show Chunk of Broken Antarctic Ice Shelf" (press release, 16 April 1997); National Snow and Ice Data Center, "Breakup of Larsen B Ice Shelf May Be Underway" (press release, 24 March 1998).

200 two hundred feet (sixty-six meters) of sea-rise: Hube, "Recent Changes to Antarctic Ice Shelves," Fig. 1.

201 famous 1978 paper: J. H. Mercer, "West Antarctic Ice Sheet and CO_2 Greenhouse Effect: A Threat of Disaster," *Nature* 271 (1978): 321–325.

201 size, description of WAIS: *Time*, 14 April 1997, 72; Charles Bentley, geophysicist, University of Wisconsin, communication with author, Madison, Wisc., 28 June 1999.

202 1970s concerns about WAIS break-up: Mercer, "West Antarctic Ice Sheet"; David Vaughan, glaciologist, British Antarctic Survey, interview by author,Washington/Cambridge, UK, 10 September 1998.

202 eighteen feet: *Time*, 14 April 1997, 76; *Science*, 21 February 1997.

202 at least 5 degrees C required: Author Interview, David Vaughan, glaciologist, British Antarctic Survey, interview by author, Washington/Cambridge, UK, 10 September 1998; Charles Bentley, geophysicist, University of Wisconsin, communication with author, Washington/Madison, Wisc., 26 June 1999. Bentley believes the figure is closer to 8 degrees C.

204 IPCC's Antarctic estimates and "zero" entries: IPCC, "Changes in Sea Level," in *Climate Change 1995*, 380.

205 *Endurance* expedition photographer James Hurley as quoted in Caroline Alexander, *The Endurance: Shackleton's Legendary Antarctic Expedition* (New York: Alfred A. Knopf, 1998), 132. Alexander's book, which is richly illustrated with Hurley's photographs, is an excellent and chilling account of the star-crossed expedition.

206 James Hurley, quoted in ibid., 181.

206 For a thorough account of the extermination of the great auk, read Mowat, *Sea of Slaughter* (Shelburne, Vt.: Chapters, 1984), 16–39.

208 Emperor Penguin adaptations: Alastair Fothergill, *A Natural History of the Antarctic* (New York: Sterling Publishing, 1995), 192–201.

208 evolution of penguins: David Campbell, *The Crystal Desert* (New York: Houghton-Mifflin, 1992), 80.

209 peninsular penguin life-cycles: Fothergill, *Natural History of the Antarctic*, 112–114; Ron Naveen, *Waiting to Fly* (New York: William Morrow, 1999), 56–57, 118–120.

210 Adélies and pack ice: William R. Fraser and Wayne Z. Trivelpiece, "Factors Controlling the Distribution of Seabirds: Winter-summer Heterogeneity in the Distribution of Adélie Penguin Populations," *Antarctic Research Series* 70 (1995): 257–272.

211 winter sea ice: *Newsweek*, 11 August 1997; *Washington Post*, 4 September 1997; William R. Fraser, Wayne Z. Trivelpiece, David Ainley, and Susan G. Trivelpiece, "Increases in Antarctic Penguin Populations: Reduced Competition with Whales or a Loss of Sea Ice Due to Environmental Warming?" *Polar Biology* 11 (1992): 525–531.

211 whaling records: *Washington Post*, 4 September 1997.

211 satellite records: Joel Comiso and Stan Jacobs, "Climate Variability in the Amundsen and Bellinghausen Seas," *Journal of Climate* 10, no. 4 (April 1997): 697.

211 Palmer Adélie declines: William Fraser and Donna Patterson, "Human Disturbance and Long-term Changes in Adélie Penguin Populations: A Natural Experiment at Palmer Station, Antarctic Peninsula," in B. Battaglia et al., eds., *Species, Structure and Survival* (Cambridge: Cambridge University Press, 1997), 445–452.

211 King George Adélie declines: Naveen, *Waiting to Fly*, 266–267.

211 Chinstrap and Gentoo increases: Bill Fraser, penguin biologist, Montana State University, interview by author, Washington/Bozeman, Mont., 4 May 1999; Fraser et al., "Increases in Antarctic Penguin Populations," 525–531.

211 seal population changes: A. W. Erickson and M. B. Hanson, "Continental Estimates and Population Trends in Antarctic Ice Seals," in K. R. Kerry and G. Hempel, eds., *Ecological Change and the Conservation of Antarctic Ecosystems* (New York: Springer, 1990), 253–264.

213 krill description, numbers, whale predation: Fothergill, *Natural History of the Antarctic*, 43–48.

213 krill life cycles from Campbell, *Crystal Desert*, 43–48, 97–107.

214 Loeb study: V. Loeb et al., "Effects of Sea-ice Extent and Krill or Salp Dominance on the Antarctic Food Web," *Nature* 387, no. 6635 (26 June 1997).

215 CFCs and ozone destruction: *Seattle Times*, 6 April 1997, A1, D12.

216 ozone hole at Pole: Patrick Neale, Smithsonian Environmental Research Center, interview by author, Edgewater, Md., 15 September 1998.

216 Arctic ozone depletion: *New Scientist*, 12 June 1999, 6; *The Christian Science Monitor*, 3 May 1999, 1; *New Scientist*, 1 May 1999, 28; *The Scotsman*, 29 April 1999, 3.

216 size of the 1998 ozone hole from information collected by NOAA's Total Ozone Mapping Satellite (TOMS) before and during the author's voyage on the *Gould*.

216 ozone and fish larvae: Deneb Karentz, biologist, University of California at San Francisco, interview by author, Washington/San Francisco, 5 October 1998.

217 Neale's study: Patrick J. Neale, Richard F. Davis, and John J. Cullen, "Interactive Effects of Ozone Depletion and Vertical Mixing on Photosynthesis of Antarctic Phytoplankton," *Nature* 392 (9 April 1998): 585–588.

221 Marr photo series: Mark Melcon, communication with author, Wiscasset, Maine/McMurdo Station, Antarctica, December 1998.

223 Torgesen colony extinctions: Fraser and Patterson, "Human Disturbance and Adélie Penguin Populations."

CHAPTER 8

226 Bartholomew I's speech was given in Los Angeles in November 1997; transcript reprinted in *New Perspectives Quarterly*, 1 January 1998.

227 quoted from Ecumenical Patriarchate, "Speech by the Most Reverend Metropolitan John of Pergamon" (address at the University of Thessaloniki, 28 September 1997). Posted at www.patriarchate.org.

229 original forty-nine Large Marine Ecosystems in Sherman et al., *Large Marine Ecosystems: Patterns, Process and Yields* (Washington, D.C.: American Association for the Advancement of Science, 1990).

230 the seventy-four LMEs from: Conservation International, "Global Marine Program Three Year Action Plan (1999–2001)," (internal document, 1999).

230 UN Law of the Sea, extent of EEZs: Michael Berrill, *The Plundered Seas* (San Francisco: Sierra Club Books), 34–35.

230 Antarctica is managed under the Antarctic Treaty system. See Colin Woodard, "Antarctica: Endless Détente," *The Bulletin of the Atomic Scientists* (January 1999): 10–12.

234 The Gully is one of Canada's first marine protected areas. See Colin Woodard, "Bottlenose Whale Could Bottleneck Canada's Gas-Pipe Route," *Christian Science Monitor*, 28 July 1997.

234 Marine Protected Areas concept: A. Charlotte de Fontaubert, David R. Downes, and Tundi S. Agardy, *Biodiversity in the Seas: Implementing the Convention on Biological Diversity in Marine and Coastal Habitats*, IUCN Environmental Policy and Law Paper No. 32 (Washington, D.C.: IUCN, 1996), 15–18; Elliott Norse, ed., *Global Marine Biological Diversity* (Washington, D.C.: Island Press, 1993), 218–220.

234 advantages of MPAs: *The News-Tribune* (Tacoma), 5 July 1998, A1.

236 artisanal fishers' worldwide catch: Elder and Pernetta, 82–83.

237 Senegal: *International Herald-Tribune*, 14 March 1997; Bahrain: *Moneyclips*, 11 February 1997; India: United Press International, 6 September 1996; *The Economist*, 18 March 1995.

237 fifty percent over-capacity in fisheries: John Fitzpatrick and Chris Newton, *Assessment of the World's Fishing Fleet 1991–1997* (Washington, D.C.: Greenpeace International, May 1998).

237 the FAO estimates that there are 3.5 million vessels, nearly two-thirds of which are "undecked," a category including the canoes, skiffs, and dories employed by artisanal fisheries. See UN-FAO, *Bulletin of Fishery Statistics – Fishery Fleet Statistics*, No. 34 (Rome: FAO, 1994); UN-FAO, *State of World Fisheries and Aquaculture 1998, Part I* (Rome: FAO, 1998); Peter Weber, *Net Loss: Fish, Jobs, and the Marine Environment*, Worldwatch Paper No. 120 (Washington, D.C.: Worldwatch Institute, July 1994), 32–33.

238 subsidies: UN-FAO, *Marine Fisheries and the Law of the Sea: A Decade of Change*, FAO Fisheries Circular No. 853 (Rome: FAO, 1993); Matteo Milazzo, *Subsidies in World Fisheries: A Reexamination*, World Bank Technical Paper No. 406 (Washington, D.C.: World Bank, April 1998); Greenpeace, "Where and How to Cut Back," in *Dead Ahead: Industrial Fishing Fleets Set Course for Disaster* (Washington, D.C.: Greenpeace, May 1998).

238 An excellent, detailed proposal for community-based fisheries management is Janice Harvey and David Coon, *Beyond the Crisis in Fisheries: A Proposal for Community-based Ecological Fisheries Management* (Fredericton, N.B.: The Conservation Council of New Brunswick, 1997). Harvey and Coon are writing about Atlantic Canada, but the system could serve as a model for other small-scale fisheries. Case studies of successful community management experiences worldwide can be found in Donald R. Leal, *Community-Run Fisheries: Avoiding the "Tragedy of the Commons"*, PERC Policy Series Issue Number PS–7 (Bozeman, Mont.: Political Economy Research Center, September 1996).

239 The drift-net ban is the result of United Nations General Assembly Resolution 44/225, a non-binding resolution that imposes a moratorium on the use of drift nets that are many miles long. It has been generally observed. Boyce Thorne-Miller, *The Living Ocean: Understanding and Protecting Marine Biodiversity* (Washington, D.C.: Island Press, 1999), 152.

239 $20–50 billion in subsidies: UN-FAO, *Marine Fisheries and the Law of the Sea*; Milazzo, *Subsidies in World Fisheries*; Greenpeace, "Where and How to Cut Back"; and Anne Platt McGinn, *Rocking*

the Boat: Conserving Fisheries and Protecting Jobs, Worldwatch Paper 142 (Washington, D.C.: Worldwatch, June 1998), 32.

240 On the development of New Brunswick aquaculture, see Inka Milewski, Janice Harvey, and Beth Buerkle, "After the Goldrush: Salmon Aquaculture in New Brunswick," in Rebecca Goldburg and Tracy Triplett, eds., *Murky Waters: Environmental Effects of Aquaculture in the US* (New York: Environmental Defense Fund, 1997), 131–152.

241 on the ISA disaster: Friends of Clayoquot Sound, *Nightmare in New Brunswick: A Salmon Farming Disease Lesson for BC* (Tofino, B.C.: Friends of Clayoquot Sound, April 1998).

241 history of concerns in Charlotte County: Milewski et al., "After the Goldrush," in *Murky Waters*, 131–152.

242 value of world aquaculture: UN-FAO, *State of World Fisheries and Aquaculture 1998, Parts 1 and 2* (Rome: FAO, 1998), UN-FAO, *State of World Fisheries and Aquaculture, 1996* (Rome: FAO, 1996). See also McGinn, *Rocking the Boat*, 47; on other species see *Christian Science Monitor*, 9 September 1998; and Department of Fisheries and Oceans, *Atlantic Aquaculture: Atlantic Char* and *Flounder* (pamphlets) (Halifax, N.S.: DFO, August 1996).

242 on aquaculture history, see Colin Woodard, "Fish Farms Get Fried for Fouling," *Christian Science Monitor*, 9 September 1998, 1.

243 shrimp farms and mangroves: Platt McGinn, *Rocking the Boat*, 50–52.

243 fish feed on farms: Rebecca Goldburg and Triplett, eds., *Murky Waters*, 8, 26–28.

244 ozone recovery: R. Monastersky, "Antarctic Ozone Hole Expands in Altitude," *Science News* (October 1997); Associated Press, 12 September 1995.

244 Kyoto targets and necessary emissions cuts: *U.S. News and World Report*, 15 December 1997; Christopher Flavin, "Last Tango in Buenos Aires," *Worldwatch* (November/December 1998): 14–15.

245 wind and solar energy's growth: Lester R. Brown et al., *Vital Signs 1999: The Environmental Trends That Are Shaping Our Future* (Washington, D.C.: Worldwatch Institute, 1999), 52–54; Colin Woodard, "Wind Power Takes Off as Major Energy Source," *The Christian Science Monitor*, 30 June 1999, 1.

248 Ross Gelbspan on subsidies: Ross Gelbspan, *The Heat Is On: The High Stakes Battle over Earth's Threatened Climate* (Reading, Mass.: Addison-Wesley, 1997), 180. In the FY 2000 U.S. budget, tax relief to the oil and gas industry was estimated at nearly $11 billion. Congressional Joint Committee on Taxation, *Estimates of Federal Tax Expenditures for Fiscal Years 1999–2003* (Washington, D.C.: Government Printing Office, 15 December 1998).

Index